Bridge Failures and Lessons Learnt

emerald
PUBLISHING

ice
Publishing

Bridge Failures and Lessons Learnt

Future-proofing to prevent disasters

Richard Fish

Published by Emerald Publishing Limited, Floor 5, Northspring, 21–23 Wellington Street, Leeds LS1 4DL.

ICE Publishing is an imprint of Emerald Publishing Limited

Other ICE Publishing titles:
Highway Bridge Management
Edited by Graham Cole and Richard Fish.
ISBN 978-0-7277-6554-3
ICE Manual of Bridge Engineering, Third edition
Edited by Gerard Parke and Nigel Hewson. ISBN 978-0-7277-6305-1
Prestressed Concrete Bridges, Second edition
Nigel Hewson. ISBN 978-0-7277-4113-4

A catalogue record for this book is available from the British Library

ISBN 978-1-83608-559-1

Cover photo: Andrea Izzotti/Shutterstock.com

Commissioning Editor: Viktoria Hartl-Vida
Content Development Editor: Cathy Sellars

Production Editor: Emma Sudderick
Typeset by: KnowledgeWorks Global Limited
Index created by David Gaskell

Contents

Foreword

As the current chair of the UK Bridges Board, I am delighted to have been asked by Richard to write the foreword to his book on bridge collapses. A hugely important subject, but often dismissed as something that only happens elsewhere in the world, this book brings to the fore the reasons why collapses happen and the precursor events that can indicate problems in the making.

I, like Richard in his former roles, am a custodian of ageing highway structures. The conflicting pressures of keeping the network open and available whilst undertaking essential, safety-critical maintenance does not get any easier and Richard is hugely experienced in balancing the technical, budgetary, customer and political demands that are placed upon bridge managers. In his current role as Technical Secretary of the Bridge Owners Forum (the research arm of the UK Bridges Board), Richard regularly reports on collapses and precursor events around the world and has given numerous presentations to the industry on the subject. He has also been instrumental in forging UK Bridges Board and Bridge Owners Forum links with CROSS (Collaborative Reporting for Safer Structures) and is exploring how the learning from UK precursor events and collapses can be shared with the wider bridge community.

Many authorities no longer have the breadth and depth of experienced bridge engineers in their employ to inspect, manage and maintain their bridge stock, yet are expected to do just that. With difficult investment decisions to be made, asset owners are under pressure to do more with less, be that to carry ever-increasing loads, reduce material use, minimise carbon and keep bridges in service for as long as possible. However, without regular maintenance to structures, this cannot go on indefinitely without restrictions or, in the worst case, a significant failure.

Precursors in the form of element failures happen regularly, a source of frustration to the public by causing inconvenient traffic disruption while a bridge joint is 'fixed' during rush hour, for example. However, often these unplanned repairs are only a sticking plaster until a larger maintenance scheme can be implemented with the underlying repair not able to be completed in an overnight possession and patched up as best as possible to allow the road to reopen. Until a permanent repair can be undertaken, joints will continue to leak, unseen concrete or steel below will continue to deteriorate and potholes will continue to form in the carriageway above the deck. These precursor events are often an indication of something that will require both significant investment and time to put right, work that should have been completed at the optimum intervention point and not left to an emergency repair.

Asset management champions the importance of understanding the condition of an asset and intervening at the optimum point to minimise whole-life cost and unplanned disruption. Of course, this makes perfect sense. However, for publicly funded asset owners, this isn't always as easy as it sounds. For bridges, it is only recently that competency-based inspector qualifications have become available, following the UK Bridges Board lead to instigate the development of such a scheme. Good-quality condition data is only the first crucial part of the story, however. In the book, Richard discusses bridge condition, the knowledge required of inspectors, engineers and managers, the many risk-based decisions that need to follow promptly and the trust that needs to be engendered with decision makers to ensure sufficient priority is given to implement essential maintenance, repair and renewal of our ageing infrastructure and ensure public safety. If decision processes are not followed through to their conclusion, there is a significant risk of element or structural failure. No bridge manager wants this to occur on 'their watch'.

Knowing when an asset will fail is not an exact science. Our models contain assumptions; the loads, the materials, the condition and the structural behaviour are all approximations. However, bridge engineers are best placed to advise on the risks when they are in possession of all the facts and can use their engineering judgement and expertise to come to an informed decision. This is where knowledge of previous failures, as set out in this book, is invaluable. Knowing where to start looking for that needle in the haystack of a network of competing priorities is essential.

To my knowledge, there is no definitive guide to bridge failures. There are many excellent reports on individual bridge collapses or failures of infrastructure, such as in the USA and Royal Commission reports on box-girder failures in the 1970s. Significant improvements were made following these latter failures with the introduction of Technical Approval and the requirement for independent checking, depending on the complexity of the structures. Technical approval and independent checking are now taken for granted in new construction but what about managing the current stock and keeping it in safe service? The number of existing structures far outweighs the number of new structures built in any year so how do we ensure that lessons are learnt when things go wrong?

This book, in my opinion, is long overdue and is essential in highlighting the criticality of bridges and the risks that we, as a society, take in using the infrastructure we implicitly rely upon. To fully understand the risks and to give opportunities to learn from others, the book sets out case studies, as well as discussing knowledge sharing, condition, vulnerable details, climate change and ethics; these are all crucial factors that bridge managers must consider in order to effectively communicate the risks to their senior managers and ultimately the political masters who decide how our valuable resources should be spent.

Richard is the perfect author for this book as a highly respected member of the UK and international bridge engineering community and commentator on bridge failures worldwide. I would like to personally thank him for his support to the UK Bridges Board, the Bridge Owners Forum and to myself over the years. Read and learn from this excellent book; I know I will.

Dr Hazel A McDonald BEng (Hons), PhD, CEng, FICE
Chief Bridge Engineer (Head of Structures), Transport Scotland
Chair of the UK Bridges Board

Preface

Bridges of all ages, structural forms and spans are vital elements in our transportation networks across the globe. They carry every imaginable mode of transport, from high-speed railways to leisurely footpaths and cycleways. And, for the most part, they function as their designers had intended; as long as they are regularly inspected and appropriately maintained.

It is impossible to put a number on the world's bridge population with any degree of accuracy, but it will be at least in the tens of millions. As far as the travelling public is concerned, the vast majority of these will be 'invisible'; it will also be taken for granted that everyone crossing over, or passing under, those bridges will be safe and free from risk. They will have an implicit trust in the designers and builders and those responsible for their maintenance and management. And this is exactly how things should be.

Very occasionally, however, for whatever reason, something goes wrong, and a bridge fails. And, although this will only apply to a tiny, almost infinitely small, percentage of the total number, the consequences are likely to be huge. That implicit trust will evaporate, communities will suffer, lives will be affected and, very sadly, in some cases they will be lost. It is a sobering thought that at the time of writing, since the turn of this century and across the world, well over 1200 people have been killed as a result of a bridge collapse.

For every bridge failure, there will be a reason; or more likely, multiple reasons. It is vital that those reasons are investigated and the knowledge shared, so that lessons can be learnt. For the most part, those lessons will be about bridge engineering, and engineers in all countries should be able to apply them in order to maintain that public trust and to keep their bridges safe. Occasionally, those investigations will find that human error has been a contributory factor. That could be anything from a simple mistake or oversight to incompetence or negligence. If the latter, this is literally criminal, and the relevant legal system should swing into action.

But even if an error was innocent and easily made, there should be no shoulder-shrugging; no 'act of God' argument. Professional engineers have an ethical obligation to serve society. Not only must they be responsible for their own actions and decisions but they must also act when they become aware of something their colleagues may, or may not, have been doing.

Although my career, in its various roles as a bridge engineer, has seen a handful of failures and a few more near misses, thankfully all have been of a minor nature, and none has resulted in injury. My interest in collapses has grown over the last 20 years or so but not in any morbid sense but rather with emotions of anger, frustration and exasperation that such things should happen.

Every engineer will have made mistakes, not least the author of these pages, but the message of learning and sharing has been part of our engineering DNA for millennia and must continue. And collective learning from mistakes is another essential principle of our profession.

One of my aims in writing this book is to aid that learning process, in terms of both engineering and ethics, by looking back at bridge failures not only historically but also in recent times. My other objective is to look forwards, to discuss what we should be doing to try to prevent collapses. It is my sincere hope that the book will make some contribution to ensuring that lessons continue to be learnt; and even, possibly, in helping to prevent a collapse; and perhaps even save a life.

Richard Fish
2024

Acknowledgements

There have been many people I have worked for, worked with and have had the pleasure to have managed who have helped me during my career, now scarily approaching the 50-year mark. Retirement beckons.

Although I could not possibly mention them all, I must acknowledge the inspiration given to me by the late Jolyon Gill (1951–2011). Working with Jolyon on the Tamar Bridge strengthening and widening was not only great fun but also occasionally a white-knuckle ride! It was Jolyon who fanned the flame of my passion for suspension bridges and later encouraged my transfer into the world of the one-man consultant, where he later joined me, and also where we collaborated on some interesting projects. A brilliant engineer, taken from us far too soon.

I would also like to acknowledge another long-serving colleague and friend, Graham Cole, a stalwart of the old County Surveyors' Society (CSS) Bridges Group and, latterly, another one-man consultant; we have worked on several projects together over the last decade or so, including co-editing *Highway Bridge Management* (ICE Publishing, 2022). I am very grateful for his kind offer to review my drafts and even more so for his constructive critique.

My thanks are also due to Dr Hazel McDonald for agreeing to write the foreword and to the Emerald Publishing editorial team, notably Cathy Sellars who guided me through the production process.

Lastly, my thanks go to my wife and family, who for reasons unknown to me, still don't get bridges.

About the author

Richard Fish BSc, CEng, FICE, FIStructE, FCIHT, MIAM, FRSA

A graduate of Southampton University, Richard began the first part of his career in the public sector. In 1976, he joined Somerset County Council on its graduate training programme, qualifying as a Chartered Civil Engineer in 1980. Most of his time in Somerset was spent on bridge design and construction.

In 1989, he moved to Cornwall County Council as County Bridge Engineer. In 1991, he joined the County Surveyors' Society (CSS) Bridges Group, taking over as Secretary a year later and holding that position until 2000. From 1991 to 2000, he also chaired the South West Area Bridge Conference.

Still in Cornwall, in 2000 he was promoted to the post of Assistant County Surveyor. Less than two years later, he was promoted again to head what had been the old County Surveyor's Department to become Director of Planning, Transportation and Estates, one of only a few local authority bridge engineers to have reached that level. Remaining in this post until 2009, he was also Chairman of both the CSS Bridges Group and the UK Bridges Board from 2005 to 2009, maintaining his commitment to bridges.

Coinciding with a major reorganisation of local government in Cornwall, Richard left the public sector and established his own consultancy business in 2009, specialising in bridge asset management and long-span suspension bridges. The latter built on his experience as project manager of the award-winning Tamar Bridge strengthening and widening project between 1994 and 2001. Over the last 15 years, he has worked for well over 30 clients in the UK and overseas in countries such as the USA, New Zealand, Turkey and Malaysia. Since 2010, he has also been the Technical Secretary of the Bridge Owners Forum, a role which has enabled him to be an active member of the UK Bridges Board.

Richard's interest in bridge collapses has grown over the years, along with a recognition of the responsibility that the bridge engineering sector, especially decision makers, has to carry. He is now widely recognised as one of the UK's authorities on the subject.

emerald PUBLISHING   ice Publishing

Richard Fish
ISBN 978-1-83608-559-1
https://doi.org/10.1108/978-1-83608-556-020251001

Chapter 1
Introduction

Bridge failures in the twenty-first century should be inexcusable and yet they still occur. No collapse can ever be considered as just one of those things (stuff happens), nor unforeseeable. Especially in the modern first world, bridges should be implicitly safe and free from any risk, at least as far as the travelling public is concerned. Collapses still happen, however, with increasing frequency. A trend which is almost certainly going to get worse before it gets better. The key issue is to learn from failure. 'Those who cannot remember the past are condemned to repeat it' (Parkinson, 1993, p. 266).

1.1.   Definitions

Before delving into the subtle differences between collapse and failure, it is best to consider some definitions in the context of this book. A bridge collapse is total, after which the bridge is unusable. While a bridge failure may also result in the same outcome, a failure of an element may be anything from, for example, the need for an expansion joint replacement to an urgent closure of the structure. While the former can be fixed in a relatively short timescale, the latter may need a significant maintenance or strengthening intervention and will either be out of use, or restricted in terms of loading, for a considerable length of time. The other distinction between collapse and failure is that the former is sudden, catastrophic and often fatal. Failure, however, as defined above, can be managed, either reactively or proactively.

Another definition worth considering is that of risk. Risk management principles define risk as the product of likelihood and consequence, and by using a relatively coarse numbering system, attention can be focused on key risks while giving others a lower priority. Although risk can never be completely eliminated, risk practitioners use the acronym ALARP: as low as reasonably practicable. This is the level of risk at which there is a degree of comfort that everything that can be done is being done to keep those risks to a minimum. A more detailed discussion on all aspects of risk is given in Chapter 8.

It would be unfair, however, to expect a member of the travelling public to analyse risk in such a way. As will be noted, their assumption will be that the risk of using, or passing under, a bridge is negligible, if not zero.

1.2.   Failure

Failure in any sector is almost inevitable. Across the range of engineering disciplines, most have the relative luxury of being able to produce, test and refine prototypes to ensure that the finished product is as risk free as possible. That process is a microcosm of that which civil and structural engineers have to follow as they develop concepts and ideas for various structural forms and materials. The fundamental difference is that the first can be conducted in a controlled environment, away from the public gaze, whereas the second is in full view and open to both scrutiny and criticism.

A failure of a small mechanical or electrical component while under development in a laboratory will help to inform decisions taken to ensure that the finished product meets all expectations. While a failure of a bridge can also help to inform the wider bridge engineering community, this only works if causes are investigated and reported, and the knowledge gleaned is made widely available.

This loop of continuous learning, applicable to all aspects of engineering, is summed up perfectly in the strapline of Henry Petroski's book, *To Engineer is Human – The Role of Failure in Successful Design* (Petroski, 1992). Or, to put it even better, as he does in a chapter title: '*Falling down is part of growing up.*'

Moving away from engineering, the worlds of commerce and business also recognise that there is value in failure. The UK entrepreneur Deborah Meadon has suggested that failure is a critical aspect of delivering successful business models (Meadon, 2022). The exact same principles of learning lessons, and sharing experiences, from failures can be applied in every walk of life: that we must learn from our mistakes.

Failure in bridges, however, is at a different level in terms of consequence. Although a failed business venture may cause financial hardship or even personal ruin, it seldom claims a life. Bridge failures, especially catastrophic collapses, are, more often than not, fatal.

Perhaps the closest parallel to a bridge is the modern airliner. Both carry large numbers of people and freight; and both defy gravity. Both also need rigorous inspection regimes and maintenance schedules. There is a difference, however, which is the quantum step between the attention given to aircraft and that given to bridges. I suggest that this is in part due to public perceptions of risk which are reflected in the priority that respective professions (and owners and operators) give to safety. The aerospace sector recognises the roles that various parties play in contributing to safety and sets a very high bar to prevent failures (Tremaud, 2010). This is a standard which the collective bridge engineering profession would do well to emulate.

All civil and structural engineers engaged in the design, construction, management and maintenance of bridges should have a broad interest in the historical development of structural forms and materials of bridges. Although many of these may be the legacy from centuries past, they still remain today as vital elements of our transport infrastructure for which bridge managers must continue to be responsible. That interest, however, should also extend to how the profession has benefited from past failures (Collings, 2008). Specific organisations must also learn from failures (Wearne, 2008). Wearne considers examples of post-war failures in the broadest spectrum of engineering sectors, from the 1956 Uskmouth turbogenerator failure to the Hatfield train derailment in 2000.

A specific engineering sector in which the consequences of failure can lead to multiple fatalities is chemical engineering. The UK's Institution of Chemical Engineers marked its centenary in 2022 by producing an online 'lessons learned database' (IChemE, 2022) in which detailed summaries of events are publicly accessible so that, hopefully, they will not be repeated.

1.3. Design life expectancy

Before delving deeper into this topic, it is worth commenting on the anticipated, or 'design', life of a bridge. The current theoretical design life in the UK is 120 years. This is nothing more than an arbitrary number which by no means should suggest that a bridge reaching this age would have to be demolished before it collapsed. Many bridges, notably masonry arches, have lasted many times this figure and have required only minimal maintenance interventions during their working lives.

While the primary intent of this book is to focus on total, catastrophic failures, it must be remembered that there are some bridge elements which will never meet the same design life as the bridge as a whole. Examples here would include bearings and expansion joints which will probably require replacement on numerous occasions during a bridge's life. The working life of these elements is very hard to accurately predict, being determined by many other factors, from vehicle numbers and weights even to the weather conditions when they were installed. Failure of other, more structural, elements could be much more serious, however, and these may be early symptoms which should be treated with some concern as precursor events, which are discussed in detail below.

1.4.   Context

My interest in bridge failures and collapses has developed as my bridge engineering career has progressed. Since 2010, it has been my privilege to have been the Technical Secretary of the Bridge Owners Forum, a role which has given me a front-row seat in our consideration of all bridge management matters, including reflections on collapses. My personal epiphany moment, however, was in 2007.

From 2005 to 2009, I had both the pleasure and responsibility of chairing the UK Bridges Board (UKBB). This was (and still is) a group of senior bridge engineers representing all bridge interests and sectors from each of the four countries in the UK, including devolved governments. The UKBB focused not only on high-level strategy and policy but also aimed to identify possible weaknesses in its collective approach to bridge management. In the same period, I also chaired the Bridges Group of the County Surveyors' Society (CSS, now the Association of Directors of Environment, Economy, Planning and Transportation (ADEPT)). This group represented all local authority bridge owners (of which there were about 200) throughout the UK. This group's interests were more practical, and it was an ideal forum for knowledge sharing and peer support.

Almost exactly halfway through my tenures of these positions, on Thursday 2nd August 2007, the BBC Radio 4 *Today* programme's eight o'clock news announced that a bridge had collapsed in Minneapolis in the USA, during the evening rush hour of the previous day. I remember expressing some immediate concern at this news but, as the morning wore on, became aware of the emerging story: not just any bridge but an interstate highway bridge, over the Mississippi River. This was the I-35W, built only 40 years previously (Figure 1.1). One hundred and eleven vehicles had been thrown into the river, resulting in 13 fatalities and another 145 injuries. More details of the I-35W collapse are given as a case study in Chapter 6.

At home that evening, I had a phone call from a *Sunday Times* journalist who wanted something to quote along the lines of: 'Well, this is America. It couldn't happen here in the UK, could it?' My view then has little changed from what it is now: well actually, yes it could.

1.5.   Precursor events

Before returning the focus to bridge failures, it is helpful to consider the theory of precursor events. These are akin to warning signs for motorists when there are possible hazards on the road ahead, or warning lights on the dashboard to show that there could be potential problems with their vehicle; they raise awareness that things may not be as straightforward, nor risk free, as they might appear.

Figure 1.1 I-35W collapse (courtesy of Mike Wills, 2007)

The initial notion of precursor events stems from the US aerospace sector and in particular the Aviation Safety Reporting System (ASRS, 2023). The model was later adopted and developed in the UK by CROSS-UK (Collaborative Reporting for Safer Structures) (Soane, 2021). The general concept is shown in Figure 1.2. This generic 'pyramid of risk' can also be transformed more specifically to consider bridge deterioration, as in Figure 1.3. In this case, however, the numbers of bridges at each level have only been illustratively quantified; there are innumerably more bridges in very good shape than those whose condition might be poor enough to fail, risking life and limb. This concept is further developed in Section 1.6 with a spectrum of deterioration as shown in Figure 1.4.

The thesis is that before any failure, there would have been other similar occurrences which should have given some level of advanced warning that things were not as they should be. Such occurrences may well not even have been on the bridge in question but possibly on another bridge of similar form, elsewhere in the country or even abroad. This gives added emphasis to the need for knowledge sharing. An example of the latter point is given in Chapter 5 in the case of the 1985 collapse of the Ynys-y-gwâs bridge.

The issue for bridge managers is, firstly, to recognise precursor events and, secondly, to be prepared to act. The former requires not only the appropriate degree of competence but also sufficient experience to understand when there is a potential problem. If those qualities are lacking, such as might well be the case with a newly appointed, inexperienced bridge manager, it is essential to know one's limitations and seek advice from one's peers.

Some recent examples of missed precursor events are among those case studies included in Chapter 6. Another relates to the collapse of the partly erected FIU footbridge in Miami, USA, in 2018. Here, precursor events were plainly visible for all to see; and yet were ignored. This collapse was the subject of a rigorous investigation by the US National Transportation Safety Board

Figure 1.2 The pyramid of risk (concept courtesy of ASRS and CROSS-UK)

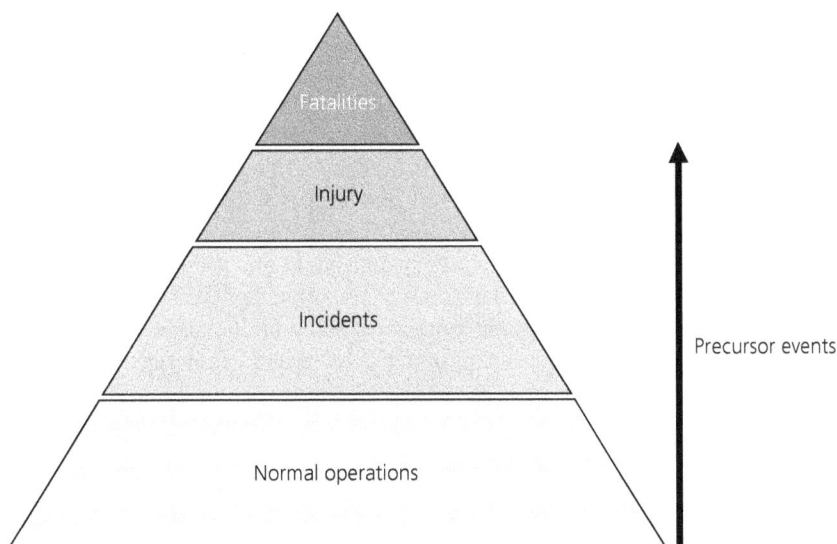

Figure 1.3 The pyramid of risk for bridges (concept courtesy of ASRS and CROSS-UK; Author's own)

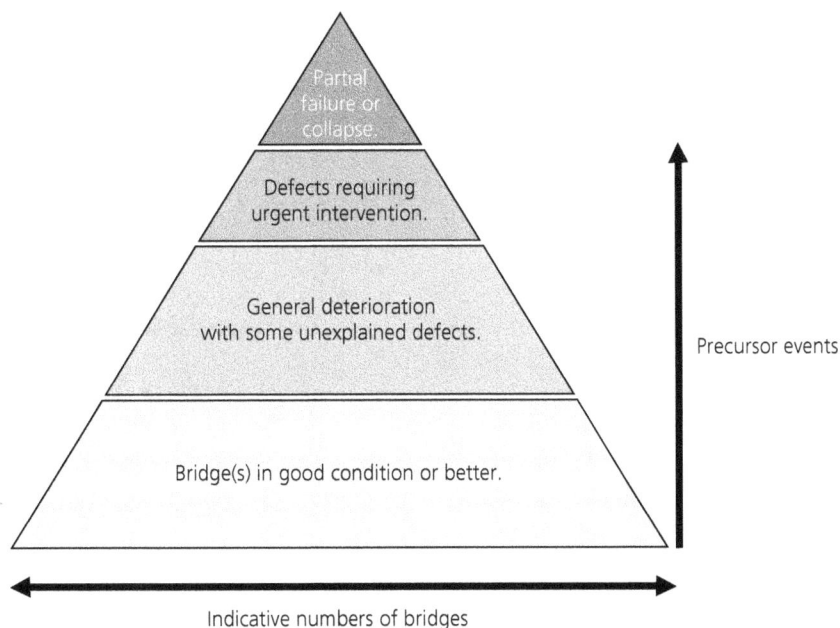

(NTSB, 2021). An excellent summary is also available from CROSS-UK (CROSS, 2020). This collapse is also covered in Chapter 8 with regard to risk, and more details of the roles of both of these investigatory organisations are given in Chapter 4.

Being prepared to act on a precursor event requires a conviction that something needs to be done and a commitment to make sure it happens. Being a bridge manager is not an easy role. As per the view of the late eminent bridge engineer, Dr Bill Harvey, there should be 'a constant low level of anxiety' for all who are charged with managing bridges (Harvey, 2021).

1.6. Perfection to disaster

Even defining what constitutes a bridge failure is not easy. At one end of a long spectrum (as shown in Figure 1.4) would be a total collapse, such as that of the I-35W bridge in 2007, leading to multiple fatalities and injuries. At the other end would be a bridge, or a bridge stock, with no apparent indication of problems. And even if a bridge manager was to find themselves in this utopian situation, there should still be no room for complacency. Indeed, in order to keep the bridge or a stock in this theoretically near-perfect condition would have required significant investment through an adequate and protected budget, a competent workforce, and a bridge inspection and management regime designed to literally maintain the status quo; to keep it in a steady state.

Figure 1.4 Bridge deterioration spectrum (Author's own)

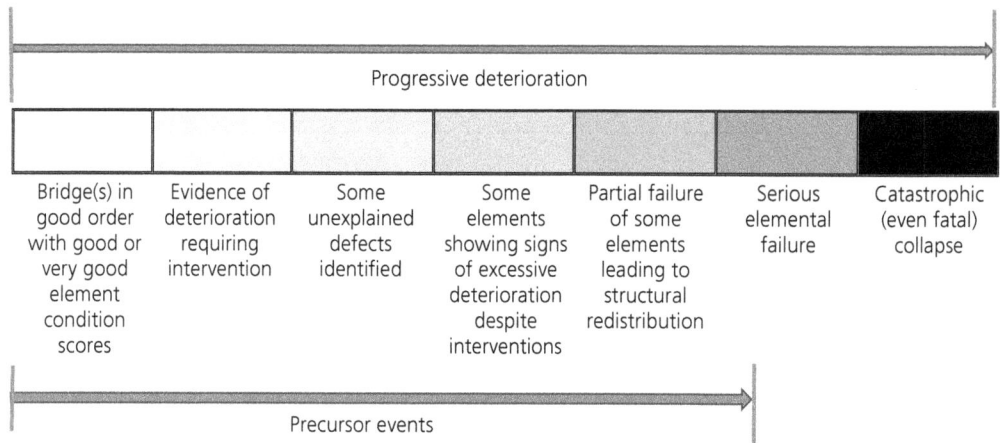

| Bridge(s) in good order with good or very good element condition scores | Evidence of deterioration requiring intervention | Some unexplained defects identified | Some elements showing signs of excessive deterioration despite interventions | Partial failure of some elements leading to structural redistribution | Serious elemental failure | Catastrophic (even fatal) collapse |

From this unlikely position of perfection, moving to the right a notch, would bring us to the real world where, inevitably, there would be some evidence of deterioration, if only through the law of entropy. That would be at a level of deterioration, however, of which the bridge manager should be fully aware; one in which their bridge management practices, built on sound, competent inspections, provide an accurate picture of the state of their bridge or stock. This would be the real world at its best, with sufficient resources available to undertake the necessary interventions to return to the steady state.

Even the next step in the spectrum also remains in the real world, where defects have been identified and accurately recorded which will cause the bridge manager some concern; to scratch their head or rub their chin. Defects which are either unexpected or not easily explained. But at least they have been identified and provide the opportunity to think hard and long about any cause-and-effect relationship.

Another step might be a case in which a planned or reactive maintenance intervention proves not to have been successful – for example, a repaired crack which reopens or continuing settlements behind an abutment. This might be a flag to those precursor events as discussed in Section 1.5.

Moving on from there would come the point of a partial failure, which could be described in risk management terms as a near miss or a close call. Any such event, however, should be a wake-up call and it is at this point that knowledge sharing should become not just desirable but essential. No bridge manager should ever work in isolation. Support networks should be in place with the appropriate level of mutual understanding that ensures that problems can be shared and advice from peers be made readily available.

The consequence of partial failures of elements depends on the structural form and levels of redundancy. It is not impossible, that even a partial failure of a critical element will trigger a total collapse, but, more probably, there will be some redistribution of load effects into adjacent members. Even the latter, however, should be seen only as a temporary state of affairs. This should lead to immediate and urgent action, such as a road closure over the bridge and taking measures to prevent access below.

Looking after any ageing infrastructure requires investment in terms of both money and people. And that is just to keep things at a steady state. Unless the right level of funding is allocated and protected for bridge maintenance, the movement to the right on the spectrum will continue. Politicians must be encouraged to reject the photo-opportunistic, ribbon-cutting, perceived-vote-winning ceremonies for new bridges, and invest in what might be considered as the less-than-sexy maintenance of the existing: the ones that potentially will kill people.

1.7.   Economic consequences

Although human life has to be the most valuable commodity, another factor that must be taken into account when considering the consequences of a bridge collapse is economics, in terms of the effects to both local and national economies. Almost by definition, the connections that bridges make between communities or, especially in more rural areas, *to* communities is immeasurable. Sever such links and there will be multiple problems, not only through an immediate economic hit but also in the subsequent impacts on deeper social issues such as accessibility, poverty and deprivation.

Highway bridges, of course, do not just carry vehicles. More so than ever before they carry utility company apparatus, from the basics, such as water, gas and electricity, through to the non-life-threatening, but ever more essential, facilities such as broadband fibre.

Whether it is connections for people or for utilities, there is seldom a quick fix – for example, when the county of Cumbria in the UK was hit with a devastating rainfall event in November 2009, the port of Workington was effectively split in two when five bridges over the river Derwent either completely collapsed or suffered damage to the extent that they had to be subject to lengthy closures, pending inspections and significant maintenance interventions. One, Northside Bridge, claimed the life of a policeman who was bravely preventing traffic from using the bridge. As well as the human tragedy, the loss of services and access had a significant and long-lasting economic impact on the community (Smyth, 2011). This flooding event, and its consequences on the Cumbria bridges, also led to evidence being given to the UK Government's Transport Select Committee (Transport Committee, House of Commons, 2010). A change of government in 2010, however, meant that no formal report was ever made public. As the primary cause of these failures was scour (Section 10.3.1), another consequence was a revised standard for scour inspections, based on a risk approach (National Highways, 2012).

1.8.   Investigations and reporting

Very sadly, bridge collapses continue to kill people. Case studies in later chapters delve into a level of detail which can only be described because there had been a rigorous investigation and, usually, an official report having been made publicly available. Such forensic scrutinies into the causes of a collapse, however, are the exception rather than the norm. There is no international standard for reporting and the quality of publicly available information ranges from excellent to zero, depending on the transport operator and/or the country in question. This topic is covered in detail in Chapter 4.

1.9.   Ethical responsibilities

Aside from the physics, and the engineering analyses, of failures, there is also the question of the behaviours of those whose fingerprints might be found on a failure. As noted above, the burden of responsibility can be significant and decisions on when or whether to intervene in a bridge's life can literally be a matter of life or death. The point here is to appreciate and accept those responsibilities and to act with integrity and honesty throughout the decision-making processes. The most important characteristic of all, however, and one worth repeating, is to know one's limitations; to know when to ask for help, when to seek advice or a second opinion, and when (and how) to escalate an issue to senior managers or politicians.

Chapter 9 discusses these ethical issues not only with respect to bridges but also by looking at other engineering sectors in which, unfortunately, there are many recent examples of unethical behaviours leading to disaster.

1.10.   Conclusion

This chapter has simply been a reasonably brief overview across the breadth of the various issues associated with bridge collapses. Other chapters drill deeper into each of the subjects that have been touched upon above. The fundamental point, however, is the need for a thorough, professional approach in all matters pertaining to bridge maintenance and bridge management.

As for the form of the book, it is split into two parts. Part 1 looks backwards at failures and collapses that have occurred over the last two millennia. This is not, however, simply a list of failures – case studies have been chosen not only to illustrate the tragic consequences but also to show where lessons have been learnt and that, as a result, both our collective engineering knowledge and public safety have been greatly improved. Part 2 is forward looking. Do we fully understand the condition of our bridges, including hidden and vulnerable details, and do we fully appreciate risk? Is our collective moral and ethical compass emphasising the burden of responsibility that bridge managers share in their day-to-day work? And are we fully aware of what the future holds, both with the added pressures of climate change and the benefits that emerging technologies might be able to provide to make our professional lives easier?

REFERENCES

ASRS (2023) ASRS – Aviation Safety Reporting System (https://asrs.arc.nasa.gov/) (accessed 20/01/2024).

Collings D (2008) Lessons from historical bridge failures. *Proceedings of the Institution of Civil Engineers – Civil Engineering* **161**(6): 20–27.

CROSS (2020) Lessons learned from the 2018 Florida bridge collapse during construction. CROSS (https://www.cross-safety.org/global/safety-information/cross-safety-alert/lessons-learned-2018-florida-bridge-collapse-during) (accessed 10/04/2024).

Harvey W (2021, January) Some thoughts on competence. *Sixty-sixth Meeting of the Bridge Owners Forum.* https://www.bridgeforum.org/bof/meetings/bof66/BOF66%20-%20Harvey%20-%20Competence%20note-202101a.pdf (accesssed 25/11/2024)

IChemE (2022) Lessons Learned Database, IChemE, https://www.icheme.org/knowledge-networks/communities/special-interest-groups/safety-and-loss-prevention/resources/lessons-learned-database/ (accessed 09/07/2024).

Meadon D (2022) Failure – There is a lot to treasure in a flop. *RSA Journal* **4**, 50.

National Highways (2012) BD 97/12 – The Assessment of Scour and Other Hydraulic Actions at Highway Structures (standardsforhighways.co.uk), National Highways, Birmingham, UK.

NTSB (2021) Pedestrian bridge collapse over SW 8th Street, Miami, Florida. https://www.ntsb.gov/investigations/Pages/HWY18MH009.aspx (accessed 10/04/2024).

Parkinson A (ed.) (1993) *The Concise Oxford Dictionary of Quotations.* Oxford University Press, Oxford, UK.

Petroski H (1992) *To Engineer Is Human.* Vintage Books, New York, USA.

Smyth M (2011) Economic and Business Recovery: Cumbria Floods Nov. 2009. https://assets.publishing.service.gov.uk/media/5a74ae6240f0b61df47779da/HistoricEnvironment-CumbriaFloods2009_0.pdf (accessed 24/11/2024)

Soane A (2021, January) CROSS update. *Sixty-sixth Meeting of the Bridge Owners Forum.* https://www.bridgeforum.org/?post_type=wpdmpro&p=489&wpdmdl=489&refresh=6540325900ab71698706009 (accessed 30/04/2024)

Transport Committee, House of Commons (2010) *The Impact of Flooding on Bridges and Other Transport Infrastructure in Cumbria.* Oral and written evidence TSO, London, UK. https://publications.parliament.uk/pa/cm200910/cmselect/cmtran/473/473.pdf (accessed 04/11/2024)

Tremaud M (2010) Identifying and utilizing precursors. *Flight Safety Foundation – European Aviation Safety Seminar – Lisbon – Mar. 15-17/2010.* https://skybrary.aero/sites/default/files/bookshelf/1442.pdf (accessed 24/11/2024)

Wearne S (2008) Organisational lessons from failures. *Proceedings of the Institution of Civil Engineers, Civil Engineering* **161**: 4–7.

Wills M (2007) *I-35W Bridge Collapse Photograph.* Licensed under the Creative Commons Attribution-Share Alike 2.0 Generic licence.

Part 1

Bridge collapses – cause and effect

Richard Fish
ISBN 978-1-83608-559-1
https://doi.org/10.1108/978-1-83608-556-020251002

Chapter 2
Historical collapses

Bridges have probably failed ever since the first spans of any significance were built. The problem, however, with regard to ancient history (and certainly in BC times), is that few collapses have ever been accurately recorded for posterity. From the early years of the first millennium AD through to the Dark Ages, very few accounts of bridge collapses can be found. From the Middle Ages onwards, however, historical records start to become more reliable and some failures from these times are noted below, either in passing or as case studies, as well as later examples up to the beginning of the twentieth century. Covering almost two millennia, this chapter can only give a flavour of the sorts of collapses that have occurred during that time. More modern collapse case studies are covered in Chapters 5 and 6.

2.1. From Rome to the Dark Ages to the Middle Ages

One of the first ever recorded collapses was a very small, but significant, part of the battle which began as Constantine reached the outskirts of Rome in 312 in order to challenge the emperor, Maxentius. The retreating Maxentius partly destroyed the masonry Ponte Milvio (Milvian Bridge, dating from 206 BC) over the river Tiber just north of the city but, in its stead, built a timber pontoon bridge in order to allow his troops to retreat and regroup, before destroying it. This plan backfired when the bridge collapsed either killing or stranding many of Maxentius' troops (Nixon and Rogers, 1994). Constantine went on to take Rome and the rest is history! Did Maxentius' mistaken strategy and his decision to rely on a temporary bridge lead to the consequence of Europe becoming a Christian continent? Probably no other collapse could ever have had such fundamental and enduring consequences.

The Romans had an unparalleled reputation as bridge builders in the first millennium AD and it was another of theirs which was one of the first recorded collapses to have claimed civilian lives. The Sint Servaasbrug over the River Meuse in Maastricht in the Netherlands (the name of the city is actually derived from the Latin for 'crossing the Meuse') was first built as a wooden bridge in AD 50. Although it is highly improbable that it was the same bridge, a collapse of what was referred to as the 'Roman' bridge occurred in 1275 while being crossed by a large religious procession. Sources record the death toll at 400 although it is likely that this figure may well have been exaggerated over the intervening years (Bredero, 1994). The present bridge, built just to the north of the failed crossing, was completed in 1298 and has gone through many transformations, including the remediation in 1948 of significant damage inflicted during World War II.

As will be seen in later chapters, extreme weather events, notably floods, have been one of the main contributory causes of bridge collapses over the centuries. In 1342, the Judith Bridge in Prague (Figure 2.1), then in Bohemia and now the capital of the Czech Republic, was subjected to a catastrophic flood. The 514-m-long multispan masonry arch bridge had been completed in 1172 by King Vladislav II, who named it after his wife who had apparently persuaded him to build it.

The flood destroyed about two-thirds of the structure and prompted the construction of a replacement, the Charles Bridge, which still stands today, although it had to be rebuilt on many occasions, also as a result of flood-induced damage (Radio Prague International, 2023).

It was, once again, crowd loading that was responsible for the next collapse example. One of the most iconic images of Venice, Italy, is the Ponte di Rialto, or Rialto Bridge, completed in 1591. The first bridge on this site dated from 1173, with others following, which included the third such structure until its collapse in 1444. A large crowd had gathered on the bridge to celebrate the marriage of the Marquis of Ferrara, and it gave way under their weight (Dupré, 1997). Although there is no record of casualties, it seems implausible that there would have been no fatalities.

2.2. Nineteenth century – UK and Europe

The nineteenth century began with the Napoleonic Wars in Europe and ended with a growth in industrialisation far beyond anyone's imagination at the turn of that century. And it was during those wars that a collapse occurred which has probably claimed more lives than any other. In 1809, Napolean's armies, led by Marshal Soult, invaded Portugal and eventually reached the city of Oporto. Although well defended on the north side of the Rio Duoro, the French broke through and forced the retreating Portuguese soldiers and civilian refugees over a wooden pontoon bridge (near the site of the present Maria Pia: Gustave Eiffel's famous crescent bridge). Out of sight of the large numbers still trying to cross, a central drawbridge span had given way and, in the ensuing panic, the crowds fleeing the French continued to force their way onto the structure. In the subsequent crush, thousands were pushed off the bridge and into the river. Although the exact number of fatalities is unknown, the widely recorded figure of 4000 is unlikely to have been greatly exaggerated (Military History Encyclopaedia, 2023).

Crowd loading was also a factor in the collapse of the Saale Bridge in Nienburg, then the territory of Anhalt-Köthen, now Germany. This was no simple bridge, but what was possibly the first stayed bridge ever built, albeit with stay-chains and not cables (Figure 2.2). Officially opened in September 1825, and designed by Christian Bandhauer, the bridge was in fact two independent structures with a relatively small-span central double-bascule section to allow high-masted vessels to negotiate a passage below (Birnstiel, 2013). The bridge's capacity had been tested (twice) with cart loads of bricks, but as part of a slightly bizarre celebration of a hare hunt by Duke Friedrich Ferdinand on 6th December 1825, crowds gathered on the new bridge. They were accompanied by a marching band which chanced to pause at midspan. The Duke was staying in a nearby castle and, in an attempt to attract his attention to the music and merriment, the crowd surged to one side. This asymmetric loading, plus some who, anecdotally, were seemingly intent on attempting to induce movement on the bridge, led to its collapse. A total of 55 souls were lost, either by drowning, or by freezing to death (Nebel, 2015).

The development of chain-supported bridges in the early nineteenth century was not unique to mainland Europe. Opened in 1826, the 44-m-span Broughton suspension bridge over the river Irwell in what is now Salford in Greater Manchester, UK, was one of the world's first chain suspension bridges. On 12th April 1831, a detachment of 74 soldiers was marching across the bridge when they found it starting to vibrate in time with their footsteps. Although the head of the column had almost crossed the bridge, there was a loud noise 'resembling an irregular discharge of firearms' and a corner of the bridge collapsed, casting some 40 soldiers into the river. Luckily at this point the river was less than 1 m deep so there were no fatalities, although about 20 were injured (Anon, 1831). A subsequent investigation found that a bolt in one of the chains had snapped. An outcome of this collapse was that the British Army thereafter ordered all soldiers to 'break step' while marching over bridges, in order for their marching not to match a bridge's natural frequency and to avoid further collapses. This order still stands to this day (Figure 2.3). While this cautionary message was also noted in France, and the same order passed to French soldiers, a similar collapse occurred in 1850 which was also partly attributable to natural frequency excitation, when the Angers Bridge collapsed, killing over 200 soldiers (Denenberg, 2009). While both the Broughton

Figure 2.2 Saale Bridge (Birnstiel, 2013)

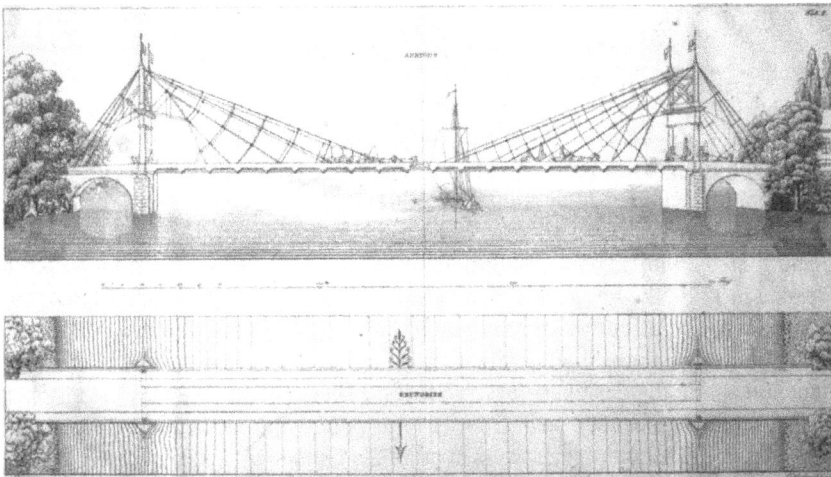

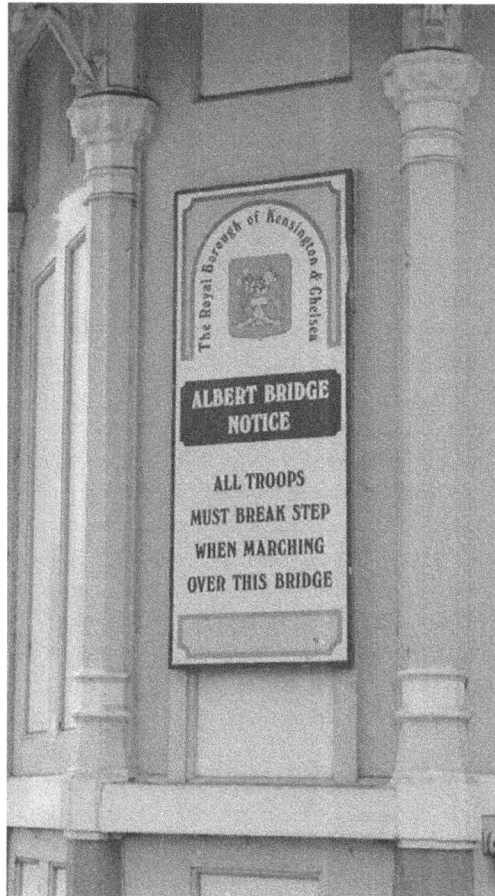

and Angers collapses were proved to be early examples of the phenomenon of pedestrian-induced vibration, fortunately the somewhat empirical contemporary engineering understanding of the problem meant that there were no more such collapses. The problem, however, still occasionally arose, most notably (and even embarrassingly) with the Millennium footbridge over the River Thames in London, which opened in 2000 (Tilly, 2011).

Another major tragedy with a chain suspension bridge occurred in Great Yarmouth, Norfolk, UK, in 1845. Opened in 1829, the bridge spanned the River Bure (Figure 2.4). Once again it appears to have been a combination of crowd and asymmetric loading that triggered the failure. There were also questions, however, about modifications made to the design during its construction, apparently without consulting the designer, Joseph Scoles. Especially notable was the not insignificant increase in the bridge's suspended span from 19 m to 26 m, while neglecting the necessary corresponding increase in the height of the towers in order to maintain the catenary profile of the suspension chains. The bridge had also been widened in 1844, with a consequent additional dead load.

Figure 2.4 Great Yarmouth bridge (courtesy of Wikimedia Commons, public domain)

The prelude to the collapse involved the promotion of a circus in the town and the unlikely stunt of a clown being towed up the river in a bathtub by four geese, before passing under the bridge. The contemporary report suggests that a large crowd had congregated against one of the parapets when an eye-bar in one of the suspension chains failed. Although the failure had been witnessed, the crowds remained on the bridge until some five minutes later, the neighbouring bar in the chain also failed and they were tipped into the river. Fatalities numbered 79 and, tragically, 59 of these were children.

The collapse was the subject of an investigation by a past president of the Institution of Civil Engineers (ICE), James Walker. A local civil engineer, William Thorold, also inspected the wreckage of the bridge and presented a report for the *ICE Proceedings* (Thorold, 1845) which was followed by a later discussion in the same publication (Rendel et al., 1845). Although noting that the design changes and widening were contributory factors, the primary cause was thought to have been the poor material in the eye-bar castings. Somewhat unbelievably, it was not until 2013 that a permanent memorial was erected near the site (Rogers, 2013).

The 1840s in the UK began a few decades of almost exponential growth in railways, with not only spectacular bridges being built but also, occasionally, some spectacular collapses. Two of the more significant of these, Dee in 1847 and Tay in 1879, are covered in detail in Chapter 5.

The expansion of the rail network in mainland Europe matched that of the UK and was also not without significant tragedies. One of the most serious of these claimed 71 lives. This was the collapse of the Münchenstein bridge in Switzerland in June 1891 (Schneider and Mase, 1968). Opened in 1875 and designed by no less than Gustave Eiffel, the bridge was a 41 m span through-truss but, as with many of Eiffel's works, it was described at the time as appearing relatively light. Many factors contributed to the collapse, not least some scour damage to an abutment which had followed an extreme flood event in 1881, and some dubious strengthening in 1890,

Figure 2.5 Münchenstein bridge collapse (Public domain, Ritter and Tetmajer, 1891)

Figure 2.5 Münchenstein bridge collapse (Public domain, Ritter and Tetmajer, 1891)

following the introduction of heavier locomotives on the network. Due to a folk-singing festival, there was also a significantly greater demand on the day of the accident, meaning that two additional coaches and a second locomotive completed the train. The bridge failed catastrophically when the two locomotives reached midspan (Figure 2.5). An inquiry was established and reported just two months later, concluding that the collapse had been caused by substandard wrought iron, the limited strengthening works in 1890, and poor workmanship dating from the original construction.

2.3.   Nineteenth century – North America

Meanwhile, on the other side of the Atlantic, there were more pioneering advances, both in the expansion of the railways and in the development of suspension bridges. Both had failures from which important lessons were to be learnt.

For the latter, it was the suspension bridge over the Ohio River at Wheeling, West Virginia, USA, which, when completed in 1849, became the longest such bridge in the world with a then-staggering main suspended span of 310 m. As an aside, it is worth noting that the design of the Wheeling bridge was a contest between two highly innovative engineers: Charles Ellet and John Roebling (the latter renowned, in particular, for his cold drawn wire and cable spinning techniques, and for the Brooklyn Bridge across the East River in New York (McCullough, 1982)), with Ellet's design being preferred. On 17th May 1854, however, the bridge deck was destroyed in strong winds causing torsional displacements and vertical undulations, allegedly reaching almost as high as the towers, although fortunately with no recorded fatalities or injuries (Kemp, 1973). The bridge went through various reconstructions, all using Ellet's towers but hindered by both the US Civil War and legal disputes. It is still in use today (Figure 2.6). The destruction of the Wheeling bridge was a salutary warning of the effect that wind loading could have on suspension bridges, as will be seen in Chapter 5.

Figure 2.6 Wheeling suspension bridge in 1949 (Public domain, courtesy of US National Park Service)

Figure 2.6 Wheeling suspension bridge in 1949 (Public domain, courtesy of US National Park Service)

Turning to railways in North America – both in the USA and in Canada – the second half of the nineteenth century saw a spate of collapses, mainly of timber trestle bridges, a popular form not least because of the availability of the material. Causes varied, but scour or, as contemporaneously reported, 'wash-out' was a particular issue, as was human error. As examples of the former, the collapse of the Springbrook Bridge in Indiana in 1859 claimed at least 41 lives (Zinuka, 2015), and the Wood River Junction bridge in 1873, another seven (Boothroyd, 2000).

With regard to human error, collapse causes took many forms. The Gasconade Bridge in Missouri collapsed in 1855, killing 35, when the preplanned inaugural train attempted to cross before works had been completed (Miner, 1972). The Sauquoit Creek Bridge in New York State killed nine people when it collapsed in 1858, simply because two trains were on the structure at the same time (New York Times, 1858). Another 14 lives were lost when the Chester rail bridge in Massachusetts collapsed in 1893 because rivets had been removed from a lattice truss girder as part of planned maintenance but without temporarily closing the bridge (Boston Daily Globe, 1893).

Bridges carrying other modes of transportation were not immune from failure. Designed for pedestrians and carriages, the Truesdell Bridge (also known as the Dixon Bridge) in Illinois collapsed in 1873. Forty-six people were killed when a large crowd, intent on witnessing a baptism ceremony, gathered against one parapet, producing such asymmetric loading that the bridge collapsed (Griggs, 2020; Figure 2.7).

Over the border in Canada, a pedestrian and tramway bridge collapsed in 1896. This was the Point Ellice Bridge over the Upper Harbour in Victoria, British Columbia, and the occasion was the celebration of Queen Victoria's 76th birthday. In order to deal with the expected crowds, the operator, the Consolidated Electric Railway Company, chose to use additional tram cars, including

Figure 2.7 Truesdell Bridge collapse, 1873 (Public domain, photo by Charles Keyes, courtesy of the Lee County Historical & Genealogical Society and the Loveland Community House, Dixon, Illinois, USA)

their heaviest, designed to carry 60 passengers. With actually a staggering 143 people on board, the bridge collapsed when the car reached midspan. The death toll was 55 (Griggs, 2021).

2.4.   Nineteenth century – Japan and India

Although most recorded collapses in the 1800s were in either Europe or North America, in bridge failure terms, the bookends of the century were further east. The first recorded fatal collapse was in Japan in 1807 when the Eitai Bridge in Tokyo (Figure 2.8), a timber structure, collapsed. This was attributed to excessive crowd loading as part of festival celebrations. Anecdotally, the number of fatalities was put at between 500 and 2000 people, although both figures seem improbable as a contemporary illustration of the collapse appears to show that it was only two of the spans which had failed (ToMuCo, 2023).

Towards the end of the century, a railway bridge over the Mysore River in India collapsed in September 1897 under a train en route to Maddur, reportedly claiming 150 lives. It appears that flooding, and presumably therefore scour, was the main contributory factor as the river was reported as being 'in flood' (Adelaide Observer, 1897).

2.5.   Conclusion

This chapter has given only a brief overview, and with only a handful of case studies, of bridge collapses through the history of the last two millennia, up to the twentieth century. The chosen examples do not purport to indicate that some are more noteworthy or tragic than others and by not mentioning some, omissions should not be taken to mean that they are not as significant. No matter what period of history, any fatal bridge failure is a tragedy for those involved and those affected. What the examples do show, however, is a variety of causes and, more often than not, the fact that it was rarely a single reason that led to the collapse. Exploring those various causes is addressed in Chapter 3.

Figure 2.8 Eitai Bridge, Tokyo (Public domain, Artist Utagawa Hiroshige, 1797–1858)

REFERENCES

Adelaide Observer (1897) Railway accident in India. https://trove.nla.gov.au/newspaper/article/162382117 (accessed 05/12/2023).

Anon (1831) Fall of the Broughton Suspension Bridge near Manchester. *The Manchester Guardian*, Manchester, UK.

Birnstiel C (2013) Collapse of a cable-stayed road bridge in Germany in 1825. *Proceedings of the Institution of Civil Engineers – Engineering History and Heritage* **166**(**4**): 207–226.

Boothroyd S (2000) Danger ahead! Historic railway disasters. http://danger-ahead.railfan.net/accidents/richmond_switch/home.html (accessed 21/11/2023).

Boston Daily Globe (1893) Blame for Chester wreck. Boston, MA, USA.

Bredero AH (1994) *Christendom and Christianity in the Middle Ages*. Eerdmans Publishing, Grand Rapids, MI, USA.

Denenberg D (2009) *Bridgemeister – 1839 Basse-Chaîne (Angers) – Angers*. https://www.bridgemeister.com/bridge.php?bid=993 (accessed 13/10/2024).

Dupré J (1997) *Bridges.* Black Dog and Leventhal Publishers, New York, USA.

Griggs F (2020) Bridge failure in Dixon, Illinois. https://www.structuremag.org/?p=16050 (accessed 21/11/2023)

Griggs F (2021) The Point Ellice Bridge failure. https://www.structuremag.org/?p=16050 (accessed 21/11/2023).

Kemp EL (1973) Ellet's contribution to the development of suspension bridges. *American Society of Civil Engineers*, **99**: 3.

McCullough D (1982) *The Great Bridge.* Simon and Schuster, New York, USA.

Military History Encyclopaedia (2023) Battle of Oporto, 29th March 1809. http://www.historyofwar.org/articles/battles_oporto_1809_1.html (accessed 27/11/2023).

Miner HC (1972) *The St. Louis-San Francisco Transcontinental Railroad; the Thirty-Fifth Parallel Project.* Lawrence, University Press of Kansas, USA, pp. 1853–1890.

Nebel B (2015) *Christian Gottfried Heinrich Bandhauer und der Einsturz der Nienburger Saalebrücke.* Books on Demand. https://buchshop.bod.de/christian-gottfried-heinrich-bandhauer-und-der-einsturz-der-nienburger-saalebruecke-bernd-nebel-9783734712050 (accessed 14/10/2024).

New York Times (1858) LATEST BY TELEGRAPH.; Eight Persons Killed and Several Fatally Injured. FORTY OR FIFTY OTHERS MAIMED. Two Trains Precipitated into Sanquoit Creek. A BRIDGE CRUSHED THROUGH. https://www.nytimes.com/1858/05/12/archives/latest-by-telegraph-eight-persons-killed-and-several-fatally.html (accessed 21/11/2023).

Nixon C and Rogers B (1994) *In Praise of Later Roman Emperors.* University of California Press, Oakland, CA, USA.

Radio Prague International (2023) History of Charles Bridge. https://archiv.radio.cz/en/static/charles-bridge/history (accessed 14/11/2023).

Rendel JM, Phipps GH, Leslie J *et al.* (1845) Discussion: The failure of the suspension bridge at Great Yarmouth. *Minutes of the Proceedings of the Institution of Civil Engineers* **4(1845)**: 293–302.

Ritter W and Tetmajer L (1891) *Bericht Über Die Mönchensteiner Brücken-Katastrophe.* Zürcher & Furrer, Zürich, Switzerland.

Rogers L (2013) Remembering lives lost in Great Yarmouth suspension bridge disaster. *Norwich Evening News*, Norwich, UK.

Schneider A and Mase A (1968) *Katastrophen auf Schienen Orell Füssli Verlag, Germany.* Translated in 1970 as *Railway Accidents of Great Britain and Europe.* David and Charles, Newton Abbot, UK.

Thorold W (1845) An account of the failure of the bridge at Great Yarmouth. *Minutes of the Proceedings of the Institution of Civil Engineers* **4(1845)**: 291–293.

Tilly G (2011) Dynamic behaviour and collapses of early suspension bridges. *Proceedings of the Institution of Civil Engineers – Bridge Engineering* **164(2)**: 75–80.

ToMuCo (2023) Tokyo Museums Collection 文化四年八月富岡八幡宮祭礼永代橋崩壊の図. https://museumcollection.tokyo/en/works/6256509/ (accessed 14/11/2023).

Zinuka J (2015) Michiana's great train wreck marks an anniversary. *South Bend Tribune.* https://eu.southbendtribune.com/story/news/local/2015/06/27/michianas-great-train-wreck-marks-an-anniversary/46616189/ (accessed 21/11/2023)

Richard Fish
ISBN 978-1-83608-559-1
https://doi.org/10.1108/978-1-83608-556-020251003
Emerald Publishing Limited: All rights reserved

Chapter 3
Causes of bridge failures

From the perspective of the layperson, or the nontechnical media (or, indeed, social media), the causes of bridge failures, and not least complete and catastrophic collapses, are often obvious and quickly identifiable. In most cases, they are also wrong. For example, when Riccardo Morandi's Polcevera Viaduct in Genoa collapsed during a thunderstorm on 14th August 2018, the immediate popular speculation was either that it had been struck by lightning or that the excessive rainwater on the carriageway had overloaded the bridge. To professional bridge engineers, both these theories are nonsense. Only through detailed forensic analyses, however, can the actual reasons be correctly identified. That said, failure can be attributed to one, or more likely several, of a relatively few distinct causes. These are discussed in this chapter.

3.1. Causes

Chapter 2 gave a brief review of a sample of failures, and their causes, from the last two millennia. Those collapses were all attributable to at least one from the list of possible causes below. Although each bridge collapse or failure is likely to be unique, it is reasonable to assume that there are just eight headline reasons:

- Errors in design
- Overstretching of engineering understanding
- Errors during construction
- Poor maintenance and management
- Unexpected material deficiency
- Errors in operation
- Environmental factors
- Force majeure

A single cause, however, would be very unusual and far more probable is a combination of two or more of these, with almost every possible permutation. An overarching additional component in almost all of these, however, is the human element. As the acknowledged founder of civil engineering, John Smeaton, identified in the eighteenth century

> *'Stone, wood and iron are wrought and put together by mechanical methods, but the greatest work is to keep right the animal part of the machinery.'* (Skempton, 1981, p. 225)

The '*animal part of the machinery*' is perhaps the most unpredictable: human beings (anyone involved in the design, construction, maintenance and management of bridges) engaged in their day-to-day work.

Before discussing these causes in more detail, it is interesting to note that Smeaton himself fully understood the consequences of a bridge collapse. His bridge over the River Tyne at Hexham, UK, only completed in 1780, collapsed '*in a sudden flood of unprecedented magnitude*' in March 1782. Smeaton not only acknowledged his error in underestimating the maximum possible flood conditions but also recognised the damage to his personal reputation

> '*It cannot now be said, that in the course of 30 years' practice and engaged in some of the most Difficult Enterprises, not one of Smeaton's works has failed. Hexham bridge is a Melancholy witness to the contrary.*' (Skempton, 1981, p. 29)

Perhaps Smeaton's comment means that the Hexham collapse was attributable to both an environmental factor *and* a design error?

Human error in all walks of life and in all sectors of engineering has continued over the last quarter of a millennium and will continue to be a factor in any incident which leads to a failure. 'People', together with 'product' and 'process', are the 3Ps mantra regularly cited by the former secretary of the Standing Committee on Structural Safety (SCOSS), John Carpenter, when identifying the three principal causes of any failure (Carpenter, 2011). People failure can either be through complacency, incompetence, negligence or even criminality. Each of these will be identified as contributory root causes elsewhere in these pages.

3.2. Errors in design

Structural design in general, and bridge design in particular, is by no means an exact science. It is based on many assumptions, from vehicle or pedestrian loading to material properties to structural actions. For this reason, design codes and standards tend to be prescriptive, with the use of partial safety factors, or their international equivalents, intended to allow for such inaccuracies. This builds in an inherent conservatism in design standards, in part to cover incorrect or misplaced assumptions.

Codes and standards, however, offer only the constraints within which a structure can be proven to be safe and in compliance with statutory requirements. The more important aspect of design is that of concept: the structural form and how it is intended to function. Conceptual design, for anything other than a simple structure, is usually a partnership between engineer and architect. And, as partnership implies, one should not impose their views on the other to the extent that the bridge fails to deliver the expected outcomes that the client had wanted. The better outcome is a structure that embraces the concept of 'structural art' (Billington, 1983), not only functioning safely as intended but also pleasing to the eye.

Bearing in mind the two principal stages of design – concept and detail – how can design errors ever account for failures? With respect to the latter and for a reasonably conventional bridge, this is less likely but, as the complexity increases, so does the designer's reliance on structural analysis software. Although sophisticated packages are widely available, they can offer a level of reassurance to the designer which may give a false sense of security (CROSS, 2022). The adage of 'rubbish in, rubbish out' is pertinent here.

No design engineer should ever be working in isolation. Procedures need to be in place for the client to approve the concept and for designs, irrespective of their complexity, to be thoroughly checked. In the UK, national standards cover this process: CG 300 – Technical Approval of Highway Structures (National Highways, 2021) for highway bridges and NR/L2/CIV/003 (Network Rail, 2012) for railway bridges. Indeed, it was a consequence of a number of highway bridge failures in the 1970s that led to this approach being adopted; these were the various steel box-girder bridge collapses which led to a committee of inquiry which reported with recommendations which were a watershed in the UK approach to design (Merrison *et al.*, 1973). Although these failures were not in any way the result of contemporary poor practice, they reflected more a lack of appreciation of potential web and diaphragm local buckling effects in the box girders. This cause of failure is addressed in Section 3.3, and a more detailed case study is covered in Chapter 5.

Perhaps the most telling of recent collapses due to design errors occurred in Miami, Florida, USA on 15th March 2018. A partly constructed footbridge, which was being built as part of an improvement to the Florida International University (FIU) campus, collapsed onto a live highway killing six people and seriously injuring another ten. Bridge collapses in the USA are subject to statutory detailed forensic investigations by the National Transportation Safety Board (NTSB) and their report into the FIU collapse (NTSB, 2019) is an excellent example of how investigations should not only be undertaken but also reported for the benefit of the wider profession. The report lists 30 findings, over half of which apportion blame to the designers who, for example, '...*made significant design errors in the determination of loads...* ' and '...*significantly overestimated the capacity of the member...* '. Not only was this collapse due to detailed design errors such as these, but it is also widely accepted that the original concept was flawed, with nonstructural embellishments such as a mast and stays, and the choice of a very unconventional truss arrangement. This example is also discussed in Chapter 8 on risk.

The issue of errors in design is not only one of technical concern but also raises the question of professional engineers' responsibility to wider society. This leads to questions of engineering ethics which are discussed in Chapter 9.

3.3. Overstretching of engineering understanding

Bridge collapses have undoubtedly occurred ever since homo sapiens began to solve the problem of using tools and materials to aid their need to travel. And, probably without realising it, early humans adopted an approach of trial and error. While this may be considered as a legitimate methodology, it should have no place in bridge design.

As engineering momentum increased, however, during the Industrial Revolution, so did the sophistication of design processes. Nonetheless, and despite some prominent Victorian engineers having undertaken extensive materials testing, there were many examples of designs which pushed the boundaries of engineering knowledge and understanding a little too far.

Examples from the Victorian era, and the contemporary rapid expansion of the railways, include the collapses of the River Dee and River Tay bridges in 1847 and 1879 respectively. Both are discussed further in Chapter 5. This trend continued into the twentieth century with the spectacular collapse of the Tacoma Narrows suspension bridge in 1940 and, as mentioned above, the steel box-girder bridge collapses of the early 1970s. These, too, are discussed in more detailed case studies in Chapter 5.

This phenomenon, of pushing the envelope of knowledge in order to deliver better designs, is not unique to the bridge, or wider structural, engineering professions. Other engineering sectors share both the same goals and the same occasional outcomes. Most notable among these is aeronautical engineering where there have been many examples of catastrophic failure. While this book is not the place to dwell on these, it is worth noting the spate of tragedies that befell the De Havilland DH 106 Comet airliner in the early 1950s (Nelson, 1993). The Comet was the first commercial jet airliner to enter service anywhere in the world and at a time when, sadly, air disasters were not that uncommon. The inquiries into the first three Comet crashes concluded, somewhat harshly in at least one case, that the main cause had been pilot error. In January 1954, however, during a flight from Rome to London, a Comet broke up in mid-air just south of the island of Elba and six crew and 29 passengers were killed. Just three months later, another Comet met a similar fate, coincidentally also over the Mediterranean, killing seven crew and 14 passengers. This led to the immediate grounding of all Comets. There followed a series of tests in which a Comet fuselage was encased in a tank of water which was pressurised and depressurised to replicate the loading endured by an actual aircraft over its working life to date, as it climbed to cruising altitude and then descended. The consequent public inquiry published its report in February 1955, finding that the principal cause had been metal fatigue caused by the cyclical loading and unloading of the pressurised cabin.

The crucial issues from the Comet examples relate not only to the detailed specifics but also to the need to share engineering knowledge. The Comet disasters proved to be a milestone in the enhancement of aeronautical engineering as well as providing a wider insight into metal fatigue, especially relating to details where stress concentrations could arise. This cross fertilisation of knowledge, especially with respect to fatigue, was in turn extremely valuable for the bridge engineering community.

3.4.   Errors during construction

When considering errors that might occur during bridge construction, it is assumed that the design process has largely been completed and that adequate works information has been issued to a contractor who will use it to build the bridge in accordance with the specification. Should a 'design and build' approach have been selected, however, then it is not unlikely that design teams will be working on site with the contractor, often providing drawings on a 'just in time' basis. While the latter has its advantages, it also carries a higher level of risk, with additional time pressures being placed on the design team, and indeed the client, to provide or accept design changes. This can occasionally lead to mistakes being made, or corners being cut, eroding any global factors of safety and possibly leading to a structural failure. Chapter 4 notes that the investigation into the Chirajara Bridge collapse in Colombia in 2018, during construction, highlighted the choice of project governance and procurement as contributory factors. Another failure during construction is that of the West Gate Bridge in Melbourne, Australia, covered in Chapter 5.

Another aspect of potential errors during construction relates to temporary works. This is an issue often overlooked by the traditional design and construction processes and yet temporary works carry a much higher level of risk and, bluntly, can kill more people than permanent works. The term 'temporary works' covers not only those structures needed to construct a bridge but also permanent works in a temporary condition, such as a deck being launched from one abutment to another. The former covers everything from scaffold access to complex launching gantries. One of the most critical pieces of temporary works, even on modest structures, is the falsework needed to support the formwork in which to cast a concrete deck. This should be designed, checked and formally approved as if it was permanent works. Historical examples of falsework failures are not uncommon, such as those in Canberra, Australia, in 2010 (Towell, 2010) and in Berkshire, UK, in 1972 when the A329(M) Loddon bridge collapsed while the deck concrete was being poured, killing three workers. The latter led to an investigation and ultimately an advisory committee on falsework and a report published in 1975, known as the Bragg Report after the Committee's Chairman (HSE, 1975).

Section 3.1 above touches on the human factor in failures; the construction process is one which relies totally on every member of the workforce fully understanding not only their personal roles and responsibilities but also how they need to work with those around them. This point in itself is totally reliant on good communication and supervision. An unsupervised worker who is unsure of what they are supposed to be doing can easily make an innocent mistake, decide on a perceived shortcut, or an alternative approach to their task, possibly leading to a serious incident or even a failure.

A further issue with construction is the degree of supervision by the designer and/or client over the contractor who is building the project. Until the turn of the century in the UK, most civil engineering projects were undertaken using forms of contract produced by the Institution of Civil Engineers, culminating in the 7th Edition in 1999 (Eggleston, 2008). In these contracts, various roles were set out, including a Resident Engineer who was required to supervise the contractor. Against a perception that these forms of contract were confrontational, a greater spirit of collaboration was encouraged by the New Engineering Contract (NEC) suite, one of its main construction contracts being the Engineering and Construction Contract. Although NEC is a suite of contracts for most eventualities (neccontract.com, 2024), for major civil engineering projects there has to be a named Supervisor. Although high-level supervision is still contractually mandated, it is the level of that supervision which is occasionally concerning, especially in the context of some suppliers or subcontractors effectively self-certifying their products based on their quality management procedures. A reduction in supervision can occasionally lead to substandard products in a bridge which may even be a latent defect to be discovered at some point after construction.

While bridge construction should be the same all over the world, there are some countries which have a poor record of bridge collapses during construction. As an example, India has seen many fatal collapses of bridges under construction, including a railway bridge near Sairang in August 2023 which killed at least 26 workers (Goksedef and Armstrong, 2023) and the spectacular collapse of the multispan Sultanganj-Aguani Ghat cable-stayed bridge over the Ganges River in June of the same year (Lateef, 2023). In fact, this was the second major collapse of the latter during its construction. Although these are the subject of ongoing investigations, it seems that there is as yet no firm commitment to make the findings public.

India, however, is by no means the only country where such collapses have occurred, but the issues of material quality, safety, supervision and quality control are likely to be factors in all cases. Of even more concern would be any evidence of corruption such as materials (for example, steel reinforcing bars) being stolen or sold, and which should have been incorporated into the bridge. This also raises the question of ethics which is considered in Chapter 9.

3.5. Poor maintenance and management

Once a bridge has been satisfactorily constructed and commissioned (normally after a maintenance period of at least one year after completion), responsibilities for its operation, maintenance and management usually pass to another party, normally the bridge owner, the client or their agent. In an ideal world, and certainly in accordance with best practice, a maintenance manual will have been prepared and issued which, as well as a full set of as-built drawings (or an equivalent BIM[1] model), should provide the owner with all the relevant data for elements and components and a recommended maintenance regime.

While the responsibility for a newly built bridge may, therefore, have shifted, the need for professional care and attention should not. In most countries, however, the emphasis still remains on building new projects, often driven by a desire to deliver political promises. And, although major maintenance – that is, in which the value of the asset is increased as a result – can be considered as capital expenditure, there is little popular recognition that this is money well spent and, hence, maintenance continues to remain the poor relation.

The bridge management process, however, whether for highway (Cole and Fish, 2022) or rail bridges (Ricketts, 2017), is one which requires a much broader skill set, and additional levels of competence, over and above those needed for more prescriptive detailed design. Bridge managers have little in the way of codes and standards to help them. The constraints of limited allocated budgets add to their burden. They have to take difficult decisions in terms of prioritisation and the timeliness of interventions in order to provide the most effective solutions.

Poor, or rather neglected, bridge management is alleged to have been instrumental in the collapse of the I-35W bridge in Minneapolis in 2007 (LePatner, 2010). It seems that there could have been a deliberate policy within the Minnesota Department of Transport to allow the condition of the bridge to deteriorate. The condition of the bridge was downgraded in 1991 from satisfactory to poor; or 'structurally deficient' in the contemporary USA designations. This was the lowest rating of bridge condition as used in the USA and bridges in this category were eligible for Federal funding for remediation. In the case of I-35W, even while in receipt of additional funding, the bridge's condition remained at structurally deficient for all of the following 16 years until it collapsed. A case study of the I-35W collapse is given in Chapter 6.

While this example may have led to catastrophic failure, there will be countless others, and for various reasons, in which maintenance interventions are delayed, postponed or cancelled. Phrases such as 'managed deterioration' or 'deferred maintenance' should never be in the bridge manager's lexicon, but being risk aware and willing to take action to prevent failure must be paramount.

3.6. Unexpected material deficiency

While a perfectly designed, well-built and well-maintained bridge might always be expected to reach its full theoretical design life, very occasionally a previously unknown phenomenon will come to light which will drastically shorten that safe working life. Such events are unforeseeable and, therefore, impossible to predict. The bridge manager, however, needs to be open-minded and adopt an attitude of 'expect the unexpected'. As and when an issue arises, it is essential that the potential problem is shared with other bridge owners.

The most germane of recent examples have arisen in concrete. Once thought to have been a completely inert material, the last decades of the twentieth century revealed instances of internal degradation within structural concrete elements. Symptoms of surface map cracking and an expansive gel within the concrete were eventually proved to have been caused by a reaction between high alkaline cements and aggregates with a high silica content such as limestone. The additional

---

[1] Building Information Modelling

component of water leads to an alkali-silica reaction (ASR) (Concrete Society, 1999). A similar reaction, with similar symptoms, and also expansive, is delayed etteringite formation (DEF) caused by early cement hydration (West, 1996).

Another issue of concrete deterioration has been found to occur when substructure concrete is exposed to sulphates in backfill, particularly in damp and cold conditions, producing thaumasite hydrate in the concrete (Wimpenny *et al.*, 2015). The effect of thaumasite reaction is an obvious reduction in concrete strength with a rapid deterioration in the material to a 'mushy' consistency.

Lastly with respect to concrete, unquantifiable damage has been inflicted on bridges which are regularly salted in freezing weather conditions for road safety reasons. Chlorides are eventually washed over or onto concrete already beginning to deteriorate through the mechanism of carbonation. When chloride salts penetrate to steel reinforcement levels, the consequent expansive reaction of the steel can either shorten a bridge's working life or require urgent intervention, both at huge expense.

It must be noted that there have been no reported catastrophic collapses due to such concrete reactions but, generally, once identified, significant interventions have had to be put in place to pre-empt possible failures. As an example, one of the most affected areas, the southwest region of the UK, had a number of ASR problems (Croke *et al.*, 2005) and several major projects had to be implemented, such as the Marsh Mills viaducts replacements in Plymouth (Whitton *et al.*, 1999).

3.7.   Errors in operation

This category principally covers changes in external structural actions, notably with respect to loading. For structural elements, such as parapets and piers, vehicle widths will also be a factor as well as, for overbridges, the consequence of impact from over-height vehicles on superstructures.

In the UK, for example, a large proportion of the bridge stock, both road and rail, dates from the eighteenth and nineteenth centuries, with the masonry arch as the predominant structural form. Live loading when those bridges were designed was negligible when compared with loads being taken today. Pressures of the modern first world have seen numerous step changes in highway design loads since the first British Standard was published in 1953 (Dawe, 2003). Similar increases have taken place around the world and translating actual loads to design loads is a complex process (OBrien *et al.*, 2022).

As loads increase, it is incumbent on the bridge manager to check that their bridges are still able to carry those actions. If not, there are two choices: either to implement permanent restrictions (on weight, or numbers of traffic lanes, or both) or to strengthen the bridge. Recognising that the latter will inevitably take some time to prepare, bridges should be managed accordingly, taking due account of risk (National Highways, 2020).

Returning to the collapse of the I-35W bridge, which again is covered in more detail as a case study in Chapter 6, there are also lessons to be learnt in terms of operational errors. Considering the fact that the bridge had been designed in 1964, just eight years after the federal decision in 1956 to embark on the national interstate highway programme, and at a time when most of the USA's freight was carried by rail, I-35W had been the subject of two significant operational interventions. Firstly in 1977, when the number of lanes was increased from six to eight and an additional concrete overlay applied and, secondly in 1998, when additional concrete median (central reserve) and side barriers were added, all of which added weight to the bridge. Another factor which should be considered is that I-35W, as a steel truss bridge, was designated, in the

American parlance, as 'fracture critical'. Although, since 2022, this term has been replaced by 'nonredundant steel tension member', both effectively mean that there is zero redundancy; failure of one member will lead to total and catastrophic collapse. That structural form was not unique to I-35W. In fact, there are at least 20 000 fracture critical bridges still in use on the USA interstate highway network.

While it is probable that these interventions were the subject of structural checks before implementation, there is little evidence of any subsequent strengthening, and the 1977 and 1998 works proved to be contributory factors to the 2007 collapse (NTSB, 2008).

Another aspect of relying on weight or width restrictions to ensure the integrity of a structure is the need to enforce them. In November 2019, a 50 t vehicle drove past a 19 t weight restriction sign and onto the suspension bridge over the river Tarn between Mirepoix and Bessières in southwest France. The bridge collapsed, killing two people.

3.8.   Environmental factors

As noted in Chapter 2, many historical bridge collapses were attributable to environmental factors, invariably with regard to river crossings and high flows. These could be due to the volume of water attempting to pass under the structure, through the impact of debris being carried along by flow, or through scour action which removes areas of the riverbed, exposing shallow foundations and leading to collapse. The last of these, scour in one form or another, is probably the biggest single cause of bridge collapses around the world.

As well as flood debris damaging a bridge through impact, an accumulation, or even a lone tree trunk, trapped against the structure can have significant effects on the hydraulics of the watercourse, with higher flow rates greatly exacerbating the scour risk (Ebrahimi et al., 2020).

Another environmental factor that can cause failure, especially in long-span and/or lightweight bridges, is wind. The collapse of the Tacoma Narrows bridge in 1940 exposed an international lack of understanding of aerodynamics and the fact that it was caught on camera has added to its infamy. This failure, and more importantly the lessons learnt from it, is the subject of a case study in Chapter 5.

In the present century, a new phenomenon is upon us which will greatly add to the risk of failure due to environmental factors: climate change, the consequences of which are covered in Chapter 10.

3.9.   Force majeure

The contractual definition of force majeure covers unforeseeable and unavoidable natural catastrophes as well as those created by humans. While some natural disasters are noted above under environmental factors, there are others such as earthquakes and tsunamis. Although design and detailing codes in earthquake-prone countries prescribe additional resilience, a more severe earthquake could lead to localised failures or even global collapses.

The obvious cause attributable to humans is open warfare. Similarly, terrorism is also a concern. The strategic nature of bridges and the vital links they provide in transport networks literally make them a target. Terrorism threat levels should be taken into account and extra vigilance is needed in protecting these key assets.

Another human factor concerns those who are going over or passing under a bridge, ranging from a vehicle impacting a parapet to a ship colliding with a bridge support. The most significant of the latter in terms of the death toll occurred when a support of the Sunshine Skyway Bridge in Florida, USA, was hit by a ship in 1980. The subsequent collapse claimed 35 lives. This is another case study considered in Chapter 5. More recently, the catastrophic collapse of the Francis Scott Key Bridge in Baltimore, USA, under similar circumstances in March 2024 will almost certainly trigger a review of ship protection systems (Pashby and Hakimian, 2024). A reflection on risk associated with this collapse is given in Chapter 8.

3.10. Combinations

As noted throughout this chapter, a single cause is unlikely to lead to a total failure. Combinations of a number of causes are far more likely. The combinations of factors can be compared to the Swiss cheese model which has been applied in numerous processes across various sectors (Reason, 1990). Figure 3.1 demonstrates the principle. If each possible cause of collapse is considered to be a risk, then it should be considered as a slice of Swiss cheese. The bigger the risk, the bigger the size and number of holes. If there is no risk, then the slice has no holes and, therefore, should be omitted from the model. This is a qualitative concept and not intended to be overanalysed but can help to give an indication of how near a structure might be to failure. A wider assessment of the approach to risk is covered in Chapter 8.

As an example, we can again revisit the I-35W. It was known that there may have been design issues, not least by the fact it was designed to 1964 standards. Although there had been an error in construction (with respect to the sizing of a gusset plate which had started to buckle well before the collapse), this had not been identified by inspections and was only recognised after the failure and as part of the NTSB investigation. Both maintenance and operational issues, however, were known to be potentially problematic. That gives four slices of cheese, each with numerous holes and, therefore, an increased risk of collapse.

Figure 3.1 Swiss cheese model (public domain, courtesy of Wikimedia Commons, licensed under the Creative Commons Attribution-Share Alike 3.0 unported licence. https://creativecommons.org/licenses/by-sa/3.0/deed.en)

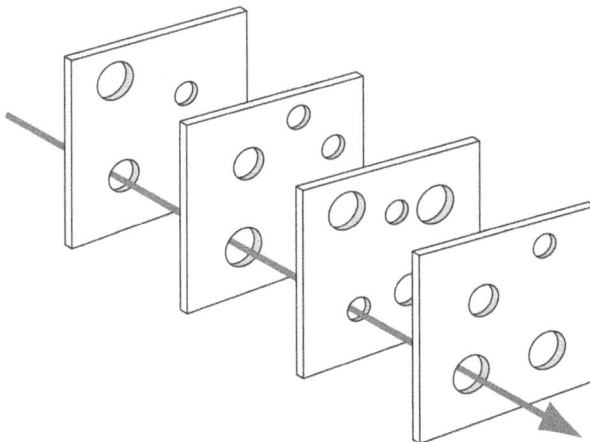

3.11. Conclusion

It is hopefully clear from this chapter that causes of failure and collapse are numerous and complex. The important point is that investigations should not only identify those causes but also, and more importantly, make the findings available so that our collective knowledge and understanding of our bridges is enhanced. This theme is further developed in subsequent chapters.

REFERENCES

Billington D (1983) *The Tower and the Bridge.* Basic Books, New York, USA.

Carpenter J (2011) Briefing: We are all risk managers. *Proceedings of the Institution of Civil Engineers – Forensic Engineering* **164(1)**: 5–5.

Cole G and Fish RJ (eds) (2022) *Highway Bridge Management.* ICE Publishing, London, UK.

Concrete Society (1999) *Technical Report 30 Alkali-Silica Reaction – Minimizing the Risk of Damage to Concrete*, 3rd edn. The Concrete Society, Sandhurst, UK.

Croke K, Pucknell B and Pearson-Kirk D (2005) Bridge management studies to reduce whole life costs. *Fourth International Conference on Current and Future Trends in Bridge Design, Construction and Maintenance.* Thomas Telford, London, UK, pp. 305–312.

CROSS (2022) *Modelling of Structures.* CROSS Safety Report 994 Modelling of structures. (https://www.cross-safety.org/uk/safety-information/cross-safety-report/modelling-structures-994 (accessed 20/01/2024)

Dawe P (2003) *Traffic Loading on Highway Bridges.* Thomas Telford, London, UK.

Ebrahimi M, Djordjevic S, Panici D and Tabor G (2020) A method for evaluating local scour depth at bridge piers due to debris accumulation. *ICE Bridge Engineering* **173(2)**: 86–99 10.1680/jbren.19.00045.

Eggleston B (2008) *The ICE Conditions of Contract*, 7th edn. Wiley, London, UK.

Goksedef E and Armstrong K (2023) India bridge collapse: At least 26 killed at construction site. *BBC News,* 23 August. https://www.bbc.co.uk/news/world-asia-india-66590539 (accessed 27/09/2023).

HSE (1975) *Final Report on the Advisory Committee on Falsework.* HMSO, London, UK. https://www.cross-safety.org/sites/default/files/2021-03/final-report-of-the-advisory-committee-on-falsework-bragg-report.pdf (accessed 22/01/2024).

ICE (2024) NEC Contracts. https://www.neccontract.com/products/contracts (accessed 30/07/2024).

Lateef S (2023) Four-lane motorway bridge collapses for second time in over a year. *The Telegraph,* 5 June. https://www.telegraph.co.uk/world-news/2023/06/05/india-motorway-bridge-collapse-river-ganges-construction/ (accessed 27/09/2023).

LePatner BB (2010) *Too Big to Fall.* Foster Publishing, New York, USA.

Merrison A, Flint A, Harper W, Horne M and Scruby G (1973) *Report of the Committee of Inquiry into the Basis of Design and Method of Erection of Steel Box-Girder Bridges.* HMSO, London, UK. https://www.istructe.org/getattachment/a1301a4d-7acb-4f3e-9623-4f8d6e1c91c1/attachment.aspx (accessed 12/01/2024)

National Highways (2020) CS 470 Management of Sub-Standard Highway Structures. *Design Manual for Roads and Bridges.* National Highways, Birmingham, UK.

National Highways (2021) CG 300 Technical Approval of Highway Structures. *Design Manual for Roads and Bridges.* National Highways, Birmingham, UK.

National Transportation Safety Board (2008) Collapse of I-35W Highway Bridge, Minneapolis, Minnesota August 1, 2007. https://www.ntsb.gov/investigations/AccidentReports/Reports/HAR0803.pdf (accessed 25/01/2024).

National Transportation Safety Board (2019) Pedestrian Bridge Collapse Over SW 8th Street Miami, Florida March 15, 2018. https://www.ntsb.gov/investigations/AccidentReports/Reports/HAR1902.pdf (accessed 25/01/2024).

Nelson TR (1993) Falling stars. *Aeroplane Monthly*, October 1993 pp. 16–18 and November 1993 pp. 20–23. IPC Magazines, London, UK.

Network Rail (2012) Engineering Assurance of Building and Civil Engineering Works. https://gat04-live-1517c8a4486c41609369c68f30c8-aa81074.divio-media.org/Corby/core-documents/nr_39.pdf (accessed 23/01/2024).

Obrien E, Nowak A and Caprani C (2022) *Bridge Traffic Loading.* CRC Press, Abingdon, UK.

Pashby T and Hakimian R (2024) Baltimore bridge collapse raises protection issues. *New Civil Engineer*, May 2024.

Reason J (1990) *The Contribution of Latent Human Failures to the Breakdown of Complex Systems.* The Royal Society, London, UK.

Ricketts N (2017) *Railway Bridge Management*, 2nd edn. ICE Publishing, London, UK.

Skempton AW (1981) *John Smeaton FRS.* Thomas Telford, London, UK.

Towell N (2010) Experts identify flaws in bridge formwork. *The Canberra Times,* 25 August. http://www.canberratimes.com.au/news/local/news/general/experts-identify-flaws-in-bridge-formwork/1922511.aspx (accessed 27/09/2023).

West G (1996) *Alkali-Aggregate Reaction in Concrete Roads and Bridges.* Thomas Telford, London, UK.

Whitton D, Rowley F and Richards A (1999) A38 Marsh Mills viaducts replacement. *ICE International Bridge Conference Proceedings.* ICE, London, UK.

Wimpenny D, Slater D, Ravindra K, Jones R and Zheng L (2015) Repair, rejuvenation and enhancement of concrete. ICE virtual library. *Thaumasite in Concrete Structures: Some UK Case Studies. Challenges of Concrete Construction: Volume 3.* https://www.icevirtuallibrary.com/doi/abs/10.1680/rraeoc.31753.0014 (accessed 12/01/2024)

Richard Fish
ISBN 978-1-83608-559-1
https://doi.org/10.1108/978-1-83608-556-020251004

Chapter 4
Investigations and knowledge sharing

The sharing of knowledge in science and engineering is a fundamental concept on the basis of which technological progress has been made over the centuries. Even the great Sir Isaac Newton had to accept, in his letter of 1675 to Robert Hooke, that '*if I have seen further, it is by standing on the shoulders of giants*' (Chen, 2003). Newton not only recognised the contributions of others but probably also understood that lessons could and should be learnt from failures as well as successes. This chapter considers some of the better practices in the investigation and reporting of bridge failures, as well as some of the not so good. While some may give only an objective account of what might have gone wrong, without drawing conclusions, others have led to a step change in engineering knowledge and understanding, resulting in a fundamental change in our approach to bridge design, construction, maintenance or management. Many such examples are given as case studies in Chapters 5 and 6.

4.1.   Introduction

It should, perhaps, come as no surprise that there is no recognised international best practice for investigation and reporting of bridge collapses. There are some exemplars, which will be discussed below, such as the National Transportation Safety Board (NTSB) in the USA and the occasional International Association for Bridge and Structural Engineering (IABSE) report. Chapters 5 and 6 contain some case studies which can only be described as comprehensively as they have been not only because of detailed investigation and reporting but also by the fact that those reports are readily accessible. Both NTSB and IABSE reports are covered in greater detail in Section 4.2.

At the other end of the spectrum, it is also the case that many collapses are never investigated and consequently no lessons can ever be learnt, with the chance that similar failures will reoccur and with the same mistakes being made. Every failure, however, should be investigated so that lessons can be learnt and shared with the wider, international bridge engineering community. There will always be various levels of interest; perhaps the primary being the legal concerns, not least in the event that litigation will inevitably follow, especially when people have been killed or seriously injured. A secondary concern will be the questions asked as to what led to the failure and the engineering reasons, such as those examined in Chapter 3. This is where candour is required. There should be a need to move aside from the legal issues and examine what had gone wrong so that engineers can work towards ensuring that any mistake or misjudgement will not be repeated in other similar situations.

There should also never be any sense of a *de minimis* scale. Even a small-span culvert that fails, if only partially and causes no injury, should be investigated and reported. How widely a report of such an instance needs to be shared is debatable but at the very least, if only an internal review is undertaken, it should be shared within peer or support groups. Not to do so, and another collapse follows with greater consequences, is not only unprofessional but also unethical. And, in extreme situations, potentially even criminal.

4.2. Best practice
4.2.1 NTSB
The USA's NTSB (NTSB, 2024) grew from its origins in the 1920s when the US Congress first set up an investigation arm as part of the Department of Commerce to investigate aircraft accidents. Although air accident investigation moved to the Civil Aeronautics Board Bureau of Aviation Safety in 1940, it was in 1967 that Congress established the NTSB as an independent agency *within* the US Department of Transportation (DOT). This was modified again in 1974, again by Congress, in order to make the NTSB truly independent, when it became a separate entity *outside* the DOT. Such independence is not only essential for objective investigation and reporting, but it also means that the USA has set a high bar as the exemplar country in this regard.

The NTSB's current remit is very broad. Investigation reports are listed under one of six categories

- aviation
- marine
- highway
- railroad
- pipeline
- hazardous materials.

Depending on the mode of transportation, bridge failures will be reported under highway or railroad incidents, and, occasionally, as in the collapse of the Sunshine Skyway Bridge in 1980 (a case study in Chapter 5) and the Francis Scott Key Bridge collapse in 2024, under the marine category.

The NTSB's website, as referenced above, provides access to all its reports, some of which have been used to summarise the examples covered in subsequent chapters.

As part of its 50th anniversary in 2017, the NTSB produced a short video (YouTube, 2017) and incorporated the following strapline, which not only sums up its own purpose but also reflects the essential point as to why bridge collapse investigation, reporting and sharing is so important

From tragedy we draw knowledge to improve the safety of us all.

4.2.2 IABSE
IABSE is not intrinsically an investigatory body but more of a broad church for international collaboration on all things associated with bridges and structures (IABSE, 2024). It can occasionally, however, be the mechanism through which an investigation is published. An example is the report into the collapse during construction of the west pylon and partially erected deck sections of the Chirajara cable-stayed bridge in Colombia in 2018. Nine construction workers were killed. Commissioned by insurance companies, this is probably the touchstone of investigation reports, published as an IABSE Case Study (IABSE, 2022). Rather than simply give a general reflection on IABSE reporting, some pertinent details of the investigation are given below.

Within just a few weeks after the collapse on 15th January 2018, a team of international experts had been assembled, undertaken an initial review, and issued an interim report on 30th May. The main objective at this stage was to determine whether the east pylon, at a similar point of construction as its neighbour, could be saved. In the event, the decision was taken to demolish it by controlled explosions.

Detailed findings of the interim and subsequent reports are brought together in the single volume of the IABSE case study. Quite rightly, no stone was left unturned, from the project's governance, procurement, planning, design and construction phases. The investigation was aided by the fact that the collapse was captured on an unrelated CCTV camera with sufficient quality to see the sequence of events on a frame-by-frame basis for the 6.9 seconds that it took from the pylon appearing to have no distress to it lying on the floor of the valley. Using design drawings and site records, a retrospective finite element model of the pylon was also analysed.

Although the IABSE report recognised that there were several shortcomings in the procurement processes, mostly with regard to the eventually selected design-and-build option, it was found that the main contributory cause to the pylon's collapse related to errors in its design, in particular an inadequate tensile capacity in the horizontal diaphragm below deck level. Furthermore, and with a somewhat chilling echo of a recommendation from investigations into steel box-girder failures almost 50 years previously (Merrison, 1973), it was noted that an independent design check had not been undertaken.

4.2.3   Confidential Reporting of Structural Safety
Initially established in the UK, Collaborative Reporting for Safer Structures (CROSS-UK) grew out of a standing committee set up between the Institution of Civil Engineers (ICE) and the Institution of Structural Engineers (IStructE) in 1976. This was the Standing Committee on Structural Safety (SCOSS) and in 1995 the two institutions were joined in their governance by the Health and Safety Executive (HSE).

In 2005, a voluntary confidential reporting system was launched with the aim of encouraging anyone in the construction, building and engineering sectors to report concerns on potentially unsafe practices with an assurance of complete anonymity. This was the Confidential Reporting of Structural Safety scheme which was the original meaning of the CROSS acronym. This process was based on that used in the USA (designed by NASA) for reporting incidents associated with aviation safety.

In 2021, following various reports (Hackitt, 2018; Hansford, 2018; Moore-Bick, 2019) into the Grenfell Tower tragedy in 2017 (a fire which consumed a residential tower block in London, UK, in which 72 people died (see also Chapter 9)), the organisation was expanded to cover fire safety and was further supported by the Institution of Fire Engineers. It was then relaunched as CROSS-UK.

International recognition of the success of CROSS-UK has led to a number of similar organisations being established in recent years: CROSS-US in the USA and CROSS-AUS covering Australia and New Zealand. At the time of writing, in Germany, CROSS-DE is being considered with a proposed association with Germany's excellent system of proof, or design review, engineers: *Bundesvereinigung der Prüfingenieure für Bautechnik e.V.* (BVPI, 2024).

By definition, CROSS-UK covers safety issues for a wide spectrum of structures but certainly not to the exclusion of bridges. Searching the website (CROSS-UK, 2024) for bridge-related reports or safety alerts is recommended at regular intervals in order to tap into this very useful vein of shared knowledge.

Although, as noted above, the HSE joined forces with ICE and IStructE on SCOSS in 1995, they are no longer a formal partner on CROSS-UK. The two organisations continue to work closely, however, and the HSE also has considerable powers both in terms of prevention through unplanned site visits and investigation of serious incidents, especially those that have led to a fatality or serious injury (HSE, 2024).

4.2.4   Bespoke investigations

Very occasionally a bridge collapse, or series of collapses, occurs of such magnitude that it rocks the profession, damages confidence, and raises concerns with the public and politicians alike. Unless the country concerned has a dedicated investigation body – such as the NTSB in the USA – there will need to be a high-level inquiry established to determine all the facts and to report its findings.

As noted above with respect to the Grenfell Tower fire, the UK has a reputation of setting up formal inquiries into such events. The criticisms, however, are, firstly, the time taken in appointing a chair, agreeing terms of reference and actually making a start, and, secondly, the length of the process itself. Such inquiries also tend to conflate legal and engineering issues. As noted elsewhere, and the NTSB represents exemplary best practice here, there should be a sense of urgency in examining the collapse from purely an engineering point of view and publishing findings, without prejudicing the more protracted legal processes which can be highly detailed in determining which, if any, parties were at fault.

Two examples of such bespoke inquiries are given as failure case studies in Chapter 5, relating to the collapse of two steel box-girder bridges while under construction within a few months of each other in 1970. Although of similar form, the bridges themselves and the circumstances of the collapses were very different, as were the consequences and the reasons. They were also on opposite sides of the planet: the Milford Haven, or Cleddau, Bridge in Wales and the West Gate Bridge in Melbourne, Australia. The former investigation concentrated on the engineering shortcomings that were associated with the failure and which eventually produced new design rules for steel box-girder bridges (Merrison, 1973). The latter took the form of a Royal Commission, the report from which provides a highly forensic review of the wide range of issues associated with the construction process (Victoria, 1971).

Although both investigations are now well over 50 years old, the findings from their reports have not aged. They are just as pertinent today and should be considered as essential reading for any engineer who might want to build their career in the bridges sector.

4.2.5   Rail Accident Investigation Branch and Air Accident Investigation Branch

The UK's Rail Accident Investigation Branch (RAIB) is an independent body affiliated to the UK Government's Department for Transport (RAIB, 2024). Its remit is to investigate and report on all significant accidents and incidents on the UK's rail network, not only on the national lines managed by Network Rail but also light rail and for the 190 heritage railways and tramways that are members of the Heritage Railway Association (HRA, 2024). It should be noted, however, that RAIB's objective is to establish causes of incidents; it has no power to apportion blame or undertake prosecutions.

Bearing in mind that RAIB's remit includes operational incidents, the numbers of their reports into structural failures are relatively few. That said, when they do report, it is with both authority and influence and bridge owners must address their findings.

The UK also has an Air Accident Investigation Branch (AAIB), almost a sister organisation to RAIB, also independent and also affiliated to the Department for Transport (AAIB, 2024). With a remit to investigate all aviation accidents, not unsurprisingly, there are no bridge failure investigations included among the many AAIB reports. The point here, however, is to demonstrate that in the UK there are governmental bodies charged with investigating incidents with transportation modes of rail and air but nothing similar with respect to highway incidents and, especially, highway bridge collapses.

4.3. Room for improvement – the UK

While the previous section has highlighted some of the best practice in the investigation of bridge failures, notably through RAIB, there are other failures in which there is only anecdotal evidence of an investigation ever having been undertaken and of which there is little or no reporting. As noted at the end of Section 4.2.5 above, there is no formalised reporting for highway bridge failures. Any investigation into such an event is entirely at the discretion of the owner of the bridge, typically the national or local highway authority. And, even if an investigation is undertaken, there is no obligation to publish a report or to share any relevant findings.

As an example, when Eastham Bridge, a Grade II listed three-span masonry arch in Tenbury Wells, Worcestershire, UK, dating from 1793, suddenly collapsed in 2016, it was a sufficiently newsworthy event to be well publicised on news websites (BBC, 2016). This was a local authority owned bridge carrying a minor road over the River Teme. Although it is understood that an internal investigation was authorised, nothing was shared even with neighbouring authorities, despite the fact that the Association of Directors of Environment, Economy, Planning and Transportation (ADEPT) has such organisational structures in place. Area bridge groups are established to provide peer support and, as necessary, to refer issues of concern to a national bridge committee (Cole and Fish, 2022). For whatever reason, it was only through a freedom of information request in 2023 that the full details could be revealed and discussed at an appropriate level (Bridge Owners Forum, 2023). Here it was revealed that the principal cause of collapse had been scour, as had long been suspected, although questions remain as to how effective the scour inspection and assessment regime had been.

In contrast, when another masonry arch, Pont Llannerch in north Wales, collapsed in 2021, also as a result of scour, the full details were shared in a matter of weeks to the same audience (Bridge Owners Forum, 2021).

The point here is that there should be no need to think that disclosing the facts associated with a bridge failure should be seen as an embarrassment or a sign of weakness. In fact, taking a 'nothing to see here' approach and not sharing the outcome of an investigation could have far more serious consequences should a similar bridge collapse under similar circumstances.

The option to be considered is whether reporting on highway bridge collapses should be either mandatory or voluntary. If the latter, then mechanisms are already in place such as those noted above with CROSS-UK. The problem arises that, unless the practice is mandatory, there is still the risk that a failure is not shared and, therefore, no lessons can be learnt. This concept is further developed in Section 4.5 below.

4.4. Other countries

It is very difficult to generalise how each country addresses the subject of investigations following a bridge failure. For the most part, investigations are legally based, often centred on the regional or local authority, and mostly intended for local consumption only. While media reports immediately after a collapse will usually state that 'an investigation is underway', or words to that effect, actually locating a formal report or outcome can be problematic, even with the benefit of the internet.

An obvious observation is that some countries are much more open than others, especially in the sense of allowing what may be perceived as sensitive information to be made widely accessible. Rather than address the pros and cons of every country with respect to their records of bridge condition, construction safety or degree of regulation, Table 4.1 gives an indication of how countries perform in these regards by listing the number of fatal collapses in this century, and the number of fatalities.

It should be remembered that this table contains only collapses that have been widely publicised. Although China is near, the top of this list, it seems probable that there have been other collapses and fatalities. Indeed, seemingly improbable figures of another 302 collapses (although not necessarily fatal) and 564 fatalities have been suggested but these are difficult to verify (Xu *et al.*, 2016).

It is also not unlikely that there will be others, especially in similar, much less open nations, which do not feature here, nor are they ever likely to. Similarly, death tolls will not always be accurate, with reports occasionally citing victims as 'missing' or 'unaccounted for'.

Some of the collapses in the table are covered as brief case studies in Chapter 6, but it is interesting to note that, thanks to the NTSB, it is the USA which provides the most detailed collapse investigations of any nation and, most importantly, they are easily accessible. It is also interesting to note that Taiwan, too, has its own Transportation Safety Board (TTSB, 2024). The country features in Table 4.1 in the 'other' category, but a case study on the collapse of the Nanfang'ao bridge in 2019 is to be found in Chapter 6.

Table 4.1 Collapses and fatalities since 2000 (up to 31st July 2024) (data courtesy of Wikimedia Commons, licensed under the Creative Commons Attribution-Share Alike 4.0 International Licence (ShareAlike 4.0 International - Creative Commons)

Country	Number of fatal collapses since 2000	Number of fatalities attributable to the collapse(s)
India	16	676
China	11	88
Guinea	1	65
Indonesia	2	62
Portugal	1	59
USA	10	55
Vietnam	1	55
Italy	3	46
Nepal	1	34
Mexico	1	26
Argentina	2	12
Canada	2	6
Others[a]	20	68
Totals	**71**	**1252**

[a] Countries that have recorded a single fatal collapse in this period in which less than 10 people died.

4.5. A sense of proportion?

In Section 4.1 above, as an introduction to this chapter, I have suggested that every failure should be reported and the information shared, no matter how small or seemingly insignificant that might be. It is not unlikely that a problem with a particular element could be a precursor to something more serious. The question arises as to whom such incidents should be reported. Similarly, how can a large number of minor failures be translated into a meaningful interpretation of a bigger issue? And how can the wider bridge community be alerted to a potentially serious problem?

This issue is not dissimilar to the DIKW theory (Ackoff, 1989). This demonstrates the need to turn vast amounts of data into usable information in order to enhance a level of knowledge which will lead to improved wisdom. This concept is shown diagrammatically in Figure 4.1.

Section 4.3 above has suggested that investigation and reporting practice in the UK needs to be improved and that the substantive point to be addressed is whether this should be mandatory or voluntary. The same section also demonstrates that there is a gap in bridge failure investigations with respect to highway bridges. While mandatory reporting may seem desirable, it is unlikely to be achieved in the short term. In order to formalise the process for voluntary reporting, CROSS-UK has developed its Voluntary Occurrence Reporting Scheme (VORS), which is loosely based on the DIKW principle and illustrated in Figure 4.2 (Bridge Owners Forum, 2024).

Figure 4.1 The DIKW principle (courtesy of researchgate.net)

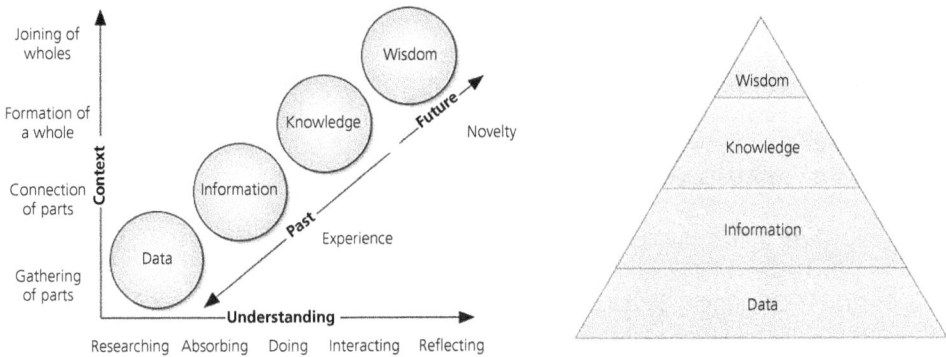

Figure 4.2 The Voluntary Occurrence Reporting Scheme (courtesy of CROSS-UK)

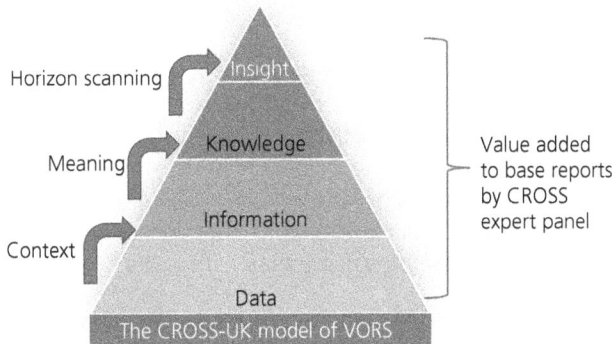

As well as providing the mechanism for failures to be reported, CROSS-UK offers to act as a repository for all such reports and to translate that data through to a level of understanding which can be more widely promulgated back to the UK bridge community.

In summary, therefore, there should be no reason for a bridge owner to say that their failure was purely a local issue and need not be reported. Indeed, the success of the VORS proposal is based on every owner making their own, possibly (to them) insignificant contribution to a wider database in order to avoid potentially major and possibly fatal failures.

4.6.   Timing

The balance which needs to be struck following a failure is between speculation about causes at a far too early stage and that of waiting many years for anything of substance to come to light. In the age of immediacy in which we live in the twenty-first century, combined with the widespread access to social media, and the inevitable consequential conspiracy theories, it is almost impossible to prevent that speculation. And now that everyone seems to have a smartphone, failures may well already be being posted on social media even before the bridge owner has been made aware.

An example of the impact of social media concerns the partial collapse of a section of wing wall and spandrel on the Plessey rail viaduct in Northumberland, UK, in October 2023. Carrying on with his late father's legacy, this collapse was featured by Hamish Harvey in his 'Bridge of the Month' series (Harvey, 2023), which not only considered the engineering reasons for the failure but also the additional pressure placed on both Network Rail and RAIB to comment when they were not, as yet, fully aware of the facts.

This is one example but there are, and will be, many others. The importance of a rapid response to a failure cannot be overstated but what is equally important is that an investigation should be subject to both the owner's policy with respect to sharing of information, and due process. This, of course, assumes that the owner has a policy. If not, this needs to be addressed as a matter of urgency.

As noted elsewhere in this chapter, in the event that a collapse is serious and that there are casualties, the situation will be handled by the police and other emergency services, closely followed in the UK by the HSE (certainly in the case that there are injuries or fatalities) or the Office of Road and Rail (ORR, 2024). In such cases, the bridge owner must be fully cooperative and allow the necessary procedures to be followed. Once those have been completed, however, they should review their own inspection and maintenance records, including any decisions (and their reasons) that may have been taken not to undertake any interventions.

4.7.   Conclusion

This chapter has revealed investigation and reporting practices of all shades: good, not so good and non-existent. It has also, hopefully, shown that there is a gold standard to which all bridge owners *should* aspire. Over and above anything else, however, it should be clear from Table 4.1 that there is a need for investigations and a sharing of knowledge. But that table also shows that the USA, with probably the best investigatory practice in the world, has still experienced a not inconsiderable number of fatal collapses this century. This, of course, is more to do with the state of the US bridge stock and the condition of many of its bridges, a subject which will be addressed in Part 2: what can be done to prevent failures rather than just simply understanding why a bridge collapsed.

REFERENCES

AAIB (2024) Air Accidents Investigation Branch – GOV.UK. https://www.gov.uk/government/organisations/air-accidents-investigation-branch (accessed 26/02/2024).

Ackoff RL (1989) From data to wisdom. *Journal of Applied Systems Analysis* **16**: 3–9.

BBC (2016) Grade II listed bridge near Tenbury Wells collapses. https://www.bbc.co.uk/news/uk-england-hereford-worcester-36373483 (accessed 26/02/2024).

Bridge Owners Forum (2021) BOF 66: Pont Llannerch. https://www.bridgeforum.org/meetings/bof66/ (accessed 26/02/2024).

Bridge Owners Forum (2023) BOF 73: Eastham. https://www.bridgeforum.org/meetings/bof73/ (accessed 26/02/2024).

Bridge Owners Forum (2024) BOF 75: CROSS Update. https://www.bridgeforum.org/meetings/bof75/ (accessed 27/02/2024).

BVPI (2024) BVPI – Bundesvereinigung der Prüfingenieure für Bautechnik e.V. https://www.bvpi.de/bvpi/en/index.php (accessed 26/02/2024).

Chen C (2003) *Mapping Scientific Frontiers.* Springer Link Home | SpringerLink.

Cole G and Fish RJ (eds) (2022) *Highway Bridge Management.* ICE Publishing, London, UK.

CROSS-UK (2024) Collaborative Reporting for Safer Structures UK (CROSS-UK). https://www.cross-safety.org/uk (accessed 26/02/2024).

Hackitt J (2018) *Building a Safer Future. Independent Review of Building Regulations and Fire Safety: Final Report.* OGL, London, UK. https://assets.publishing.service.gov.uk/media/5afc50c840f0b622e4844ab4/Building_a_Safer_Future_-_web.pdf (accessed 26/02/2024).

Hansford P (2018) *In Plain Sight: Assuring the Whole Life Safety of Infrastructure.* Institution of Civil Engineers, London, UK. https://www.ice.org.uk/media/bj0mnx4h/in-plain-sight.pdf (accessed 26/09/2024).

Harvey H (2023) Bridge of the Month 153. https://www.billharveyassociates.com/bom/153-plessey-viaduct?utm_source=drip&utm_medium=email&utm_campaign=bom153 (accessed 27/02/2024).

HRA (2024) Home – Heritage Railway Association. https://www.hra.uk.com/ (accessed 26/02/2024).

HSE (2024) HSE: Information about health and safety at work. https://www.hse.gov.uk/ (accessed 26/02/2024).

IABSE (2022) *Investigation of the Chirajara Bridge Collapse. IABSE Bulletins Case Studies CS3. Structurae.* IABSE, Berlin, Germany.

IABSE (2024) IABSE – Home. https://www.iabse.org/ (accessed 26/02/2024).

Merrison AW (1973) *Report of the Committee of Inquiry into the Basis of Design and Method of Erection of Steel Box-Girder Bridges.* HMSO, London, UK. https://catalogue.nla.gov.au/catalog/1085555 (accessed 28/01/2024).

Moore-Bick M (2019) Phase 1 Report of the Grenfell Tower Inquiry. https://www.grenfelltowerinquiry.org.uk/phase-1-report (accessed 26/02/2024).

NTSB (2024) Home. https://www.ntsb.gov/Pages/home.aspx (accessed 10/02/2024).

ORR (2024) Home – Office of Rail and Road. https://www.orr.gov.uk/ (accessed 07/02/2024).

RAIB (2024) Rail Accident Investigation Branch – GOV.UK. https://www.orr.gov.uk/ (accessed 26/02/2024).

TTSB (2024) 國家運輸安全調查委員會全球資訊網. https://www.ttsb.gov.tw/ (accessed 27/02/2024).

Victoria (1971) *Report of the Royal Commission of Inquiry into the Failure of West Gate Bridge.* CH Rixon, Melbourne, Australia.

Xu F, Zhang M, Wang L and Zhang J (2016) Recent highway bridge collapses in China: Review and discussion. *Journal of Performance of Constructed Facilities* **30(5)**. https://ascelibrary. org/doi/10.1061/%28ASCE%29CF.1943-5509.0000884 (accessed 02/06/2024).

YouTube (2017) NTSB 50th Anniversary – YouTube. https://www.youtube.com/watch?v=ux_tijGI5kw (accessed 10/02/2024).

Richard Fish
ISBN 978-1-83608-559-1
https://doi.org/10.1108/978-1-83608-556-020251005

Chapter 5
Historical case studies

Although earlier chapters have touched on many historical catastrophic collapses, they have only been addressed in brief. The following case studies are examples from the nineteenth and twentieth centuries. They are collapses which have either been the subject of a rigorous investigation or have made a significant contribution to the advancement of bridge engineering understanding; or both. The examples chosen are by no means exclusive, but each should serve to illustrate the level of structural understanding before the failure, and the consequences in terms of changes in engineering knowledge, contractual relationships or standards thereafter.

They are listed here in chronological order, not in any way as an indication of their significance. They are also simply summaries and, therefore, cannot include every detail. For a more comprehensive account, where appropriate, the reader should refer to the reference for each which has been cited at the beginning of some case studies.

5.1.   Dee – 1847
5.1.1   Context
A definitive reference here is Peter Lewis's book, *Disaster on the Dee* (Lewis, 2007).

The growth of the Victorian railway network in the UK was at its highest in the 1840s. Unlike the twentieth century motorway and trunk road building programmes, however, there had been no national strategic masterplan, but rather a large number of private railway companies promoting rail links often driven by a local demand, rather than a national strategic perspective. There were some exceptions to the principle of local ahead of national interests when the government recognised a wider benefit and, in some cases, would offer financial support. One of these, because of the opportunity it afforded to enhance the connectivity between Great Britain and Ireland, was the Chester to Holyhead Railway.

The planning process for new railways required an Act of Parliament to obtain approval for the venture and, as part of the preparation of the Bill, the promoting company would have commissioned an engineer to work up the design which would eventually be proposed and approved. In the majority of cases, the same engineer would have been retained to develop the details of the project and to supervise its construction. It is important to recall, however, in our current world of prescriptive codes and standards, that the scheme's engineer would have had absolute freedom in terms of their designs and details. It must also be recognised that the engineering profession at this time was held in such high regard that the only expectations were ones of success and absolute confidence (Petroski, 1994). The first major failure to cast a shadow of doubt on the infallibility of engineers in the UK came when the Dee Bridge collapsed on 24th May 1847, with the loss of five lives, only seven months after it had opened.

Before discussing the details of the bridge and its failure, another piece of context concerns the available materials that engineers were using to deliver their conceptual designs as economically

45

as possible. Since the opening of the spectacular single-span iron arch over the River Severn at Coalbrookdale in 1779, iron had been the preferred material for all bridges, other than masonry arches or viaducts (Cossons and Trinder, 1979). Iron came in two forms: cast and wrought. Both had been used at a relatively small scale for millennia and both forms had their strengths and weaknesses. Wrought iron gave improved ductility and had been the material of choice for bridge construction once rolled plates and angle sections could be produced to construct rivetted plate girders. Cast iron is more durable, strong in compression but weak in tension. It is also a brittle material, susceptible to sudden failure. Indeed, there had been several failures of cast-iron elements, including the collapse of a mill roof in Oldham in 1844, which claimed 20 lives.

5.1.2   The Chester to Holyhead Railway

A railway between London and Manchester, by way of Birmingham, had been completed by 1838. The city of Chester was an important hub, with other branch lines planned to and from it. It was the obvious choice from where to drive a line along the north Wales coast to the port of Holyhead on the island of Anglesey. Such a line would require significant bridge crossings of the Menai Strait and the River Conwy; this may have been a consideration in the railway company commissioning one of the top contemporary railway engineers, Robert Stephenson.

5.1.3   The Dee Bridge

In relation to the much wider crossings of the Menai and Conwy, with spans of 213 m and 122 m respectively, the bridge required to span the River Dee just outside Chester was relatively modest. The total width of the river from bank to bank on the line of the railway was 76 m but, with concerns over potentially troublesome foundations, Stephenson shied away from a masonry arch viaduct in favour of a three-span iron girder bridge with two masonry piers in the river. The arrangement was three equal spans, each of 33.2 m.

These spans were record breaking in terms of the use of cast-iron girders as hitherto a length of about 9 m (Scheer, 2010) had been considered optimum, not only from a structural engineering perspective but also in terms of the limitations of being able to cast anything longer. Stephenson overcame the former by introducing wrought-iron tie rods to effectively form a shallow truss, on a profile not dissimilar to a post-tensioning strand in a prestressed concrete deck. To achieve a beam of this length, Stephenson designed each girder to be made of three castings, joined by splices (Åkesson, 2008).

While this arrangement for a single beam may have been structurally acceptable when considered under direct loading, the detailed positioning of the transverse beams meant that they were supported on the bottom flanges of the main girders. This meant that the loading on the primary members was asymmetric, with consequential torsional effects which appear not to have been considered in Stephenson's design.

5.1.4   24th May 1847

During the morning of the fateful day, six trains crossed the bridge without incident. Ironically, Stephenson had also been on site that morning as he was worried about the timbers over the transverse members being a potential fire hazard from sparks and ashes from the steam locomotives. Such an issue had recently led to the destruction of a bridge elsewhere in the country. In an attempt to allay his concerns, Stephenson instructed a local contractor to add some 125 mm of additional ballast over the deck timbers, perhaps inadvertently adding an extra nine tons of superimposed dead load, as well as exacerbating the already asymmetric loading, on each girder.

Following these works, the 6.15 pm train (weighing some 60 t) left Chester station heading towards the bridge. As it crossed the final span, one of the main girders failed and the carriages fell into the river. The locomotive and tender, however, were just able to reach the far side, before derailing. Four passengers and the stoker were killed. Figure 5.1 is from *The Illustrated London News* of 12th June 1847, in which the cast/wrought-iron beam detail can be seen.

5.1.5   Investigation

Dee was by no means the first incident in the Victorian railway era and, as a result of others, the UK Government had established a Railway Inspectorate in 1840 through the Regulation of the Railways Act of that same year. As well as more mundane issues, such as requiring railway companies to provide data on their operations, the Inspectorate had been given powers to undertake independent investigations into accidents and to publish reports. The team of inspectors had largely been recruited from the British Army's Royal Engineers and hence had a good working knowledge of engineering principles.

The Inspectorate submitted its report about three weeks after the incident. By this time, however, a formal inquest was already underway and, as was the practice at that time, evidence was considered by a jury of local ratepayers. Although various witnesses observed not only vertical deflections in the beams but also rotational displacements, Stephenson himself presented a classic 'cause or effect?' argument, claiming that it was the train derailing, hitting the edge of the bridge and causing the beam to fracture, rather than vice versa.

The jury concluded that cast iron was not a reliable material for long-span beams and that the use of wrought-iron tension bars had been ineffective. Although the coroner had directed that negligence

Figure 5.1 River Dee Bridge collapse (courtesy of ICE Library)

380   THE ILLUSTRATED LONDON NEWS.   [June 12, 1847.

THE LATE RAILWAY ACCIDENT, AT CHESTER.

on the part of Stephenson should *not* be part of the jury's summary, they did not demur from their criticism of the design. The jury also recommended that a broader review of the use of iron in railway bridges should be undertaken and consequently a Royal Commission was established, part of which was to include some detailed testing of both cast and wrought iron. The Commission reported in 1849, by which time Stephenson had become an MP, representing Whitby.

From a modern perspective, failure of the Dee Bridge was almost certainly due to torsional buckling instability due to both an unrestrained top flange and asymmetric loading. Other theories have suggested that this was an early indication of metal fatigue, although the time between the opening of the bridge and its collapse would seem to suggest that this phenomenon would only have made a marginal contribution.

Despite the Dee Bridge collapse, Stephenson's reputation seemed to have escaped relatively untarnished, and he remains a well-respected engineer from the Victorian railway era, with many notable achievements to his name, not least his wrought-iron boxes of the longer crossings of the Chester to Holyhead line: the Conwy and Brittania bridges.

5.2. Tay – 1879

5.2.1 Context

As well as other references noted in the text below, one of the best books on this case study is probably David Swinfen's *The Fall of the Tay Bridge* (Swinfen, 2016). Professor Roland Paxton states in his Foreword that it 'deserves to be regarded as the final word on the subject'.

The previous case study reflected on the growth of the railway network in Victorian Britain, often referred to as 'railway mania'. Even after the rapid expansion of networks in the 1840s, all promoted by private companies, there emerged a rivalry between two of them to drive lines to the cities of Dundee and Aberdeen in northern Scotland (McKean, 2006).

Extending from the west coast lines to Carlisle was the Caledonian Railway Company and developing a continuation of the line from Berwick-on-Tweed to Edinburgh, to the east of the country, was the North British Railway. There was almost literally a race between them to secure the first fares to these destinations from English cities already well connected across the country. For the North British Railway, there were two very substantial engineering challenges: spanning the Firth of Forth (west of Edinburgh) and the Firth of Tay (south of Dundee). A glance at a map of east Scotland will show that these are not insignificant; both would require major estuarial crossings.

5.2.2 Thomas Bouch

Thomas Bouch was the engineer for the North British Railway. His initial proposal was a floating railway to cross both estuaries (Swinfen, 2016). This was essentially a ferry with rails on the deck onto which rolling stock could manoeuvre, and then be carried across the water to rejoin the railway on the other bank. Steam ferries were commissioned and operated this arrangement, starting across the Tay in 1851, until the new bridge opened in 1878.

Once his ferry scheme was up and running, Bouch left the company to work as a consulting engineer on other railways in the north of England and southeast Scotland. His burgeoning reputation was largely based on lightweight, economical designs for bridges and viaducts. One such viaduct was for the North Eastern Railway on a line from Barnard Castle in County Durham westwards towards Cumberland. This was the Belah Viaduct, over 300 m in length and formed of 16 spans supported on braced hollow cast-iron columns up to 30 m high. The twin tracks were carried on wrought-iron latticework girders, a form which was not only straightforward to construct but also thought, because of its many openings, to offer minimal resistance to wind loading. It was opened in 1861 and was still in use over 100 years later.

In 1862 the North British Railway initially invited Bouch to design a bridge over the Firth of Forth but two years later he was also tasked with designing the Tay Bridge.

5.2.3 The Tay Bridge

Bouch's initial design was for a 63-span crossing at a maximum height above water of over 30 m. Despite the scale of the construction, Bouch is alleged to have convinced a public meeting that his bridge would be 'a very ordinary undertaking'. To meet various concerns, however, the line of the bridge had to be altered and the number of spans increased to 80. By the time that the necessary Act of Parliament received royal assent in 1870, the number of spans had increased to 85 with the maximum being almost 75 m long. Most spans were supported on trusses below rail level. The central spans – the 'high girders' – were through trusses with the rolling stock passing inside the truss sections.

Bearing in mind that the bridge was funded by investment from shareholders, there was an obvious desire to keep capital costs as low as possible. This may have been a consideration in commissioning Bouch in the first place, as the concept of his design appears to have been modelled very closely on that of the Belah Viaduct. Similarly, after some problems with those initially awarded the contract, the contractor was eventually the same one as Bouch had employed on Belah.

Another factor, probably driven by budgetary issues, was the fact that the bridge was to carry only a single track. Well before construction had started, this was something which had attracted considerable criticism not only on the grounds of train scheduling but also with regard to concerns over the perceived slenderness of the bridge, with a local newspaper likening it to a 'washing line'.

Work began in 1871, although by that time there had been a change of the contractor (Butler and Pitts) which had initially submitted the lowest tender. This was not the only issue with the builders as another change was needed in 1873. The second contractor had been Charles de Bergue but, when he died in that year, it became clear that his company was already in financial difficulties. Bouch himself was very active in trying to resolve these issues and eventually the firm of Hopkins and Gilkes was appointed. It was Gilkes who had worked with Bouch on the Belah Viaduct. This was not the most auspicious beginning for the construction of what was to be the longest bridge in the world.

The bridge construction used many innovative techniques, now commonplace in long-span bridge building, including pneumatic caissons for the deeper foundations, together with the use of steam-driven vacuum pumps to remove silt. Individual spans were also prefabricated and moved to their final locations by barge before being slowly lifted into place as their supports were erected. Sadly, although not unusual for the time, the bridge's construction had its fair share of accidents and incidents, many of which claimed lives (a total of 20 of the workforce were killed). Ominously, some of these reflected the exposed location and the formidable weather conditions which led to a number of workers literally being blown off the bridge.

On 26th September 1877, however, the first train crossed the bridge. This inaugural crossing carried the directors of the North British Railway and their guests, but this was eight months before

Figure 5.2 The completed Tay Bridge (Public domain, courtesy of National Library of Scotland/ Wikimedia Commons)

the formal opening in May 1878. The interim period had seen the formal Board of Trade inspection of the structure, conducted by the former military engineer, Charles Hutchinson. Although he generally accepted the safety of the bridge, his prophetic caveat was that he would have liked to have observed the effects of a high wind when a train was running over the bridge.

The official opening took place on 31st May 1878 and a year or so later Queen Victoria herself crossed the bridge in the royal train on her return to London from Balmoral. In recognition of this amazing feat of engineering, Thomas Bouch was knighted at Windsor Castle on 27th June 1879 (Figure 5.2).

5.2.4   28th December 1879

In one of the worst storms for many years (during which at least five ships in the area had either sunk or run aground), at about 7.15 pm, the 13 central spans of the Tay bridge – the 'high girders' – collapsed into the Firth of Tay, together with the northbound 17.20 train from Burntisland and all its crew and passengers. There were no survivors. Although early estimates put the total number of fatalities at about 300, this was much exaggerated and later reduced to 75. More detailed examination set the true, although still shocking, figure at 59. Only 46 bodies were ever recovered (Figure 5.3).

5.2.5   The investigation

With an unlikely haste by today's standards, a court of inquiry, with just three members, including the then president of the Institution of Civil Engineers, William Barlow, was established by the Board of Trade. It convened in Dundee and first sat on Saturday 3rd January 1880, less than a week after the collapse. Evidence was given by interested parties, including Bouch, as well as witnesses to events before and after the collapse, with the latter including divers who had both examined the wreckage in the river and assisted in the recovery of some personal effects.

The court reconvened on 26th February 1880 and heard additional testimony from many parties with a variety of allegations, such as trains crossing at excessive speeds, substandard cast iron (and subsequently identified cracks) in the columns and the level of inspection and maintenance of the bridge since opening. It formally reported to Parliament in June 1880, but with two separate

Figure 5.3 River Tay after the bridge collapse (courtesy of National Library of Scotland/Wikimedia Commons. Public domain)

reports. Barlow and Colonel Yolland, the Chief Inspector of Railways, produced an objective, detailed and factual account of the collapse without apportioning blame. Their colleague, however, Henry Rothery, the Commissioner for Wrecks, presented his own version which apparently had not been sanctioned by Barlow and Yolland. His report was highly critical of many individuals, not least Bouch.

5.2.6   The cause

As with the Dee collapse, arguments of cause or effect were mooted. Did the speed of the train, combined with the force of the wind, lead to a derailment which triggered the collapse? Or were the lateral forces from the wind too great for the slender columns and their cross bracing which led to a global failure. In a similar manner as had been Stephenson's defence at Dee, Bouch supported the former argument. The answer, however, was probably a combination of the latter together with the construction of the columns – both in terms of the quality of the cast iron and the sizing of the wrought-iron bracing members.

The collapse marked not only the end of Bouch's career but also his life. Despite having been damned by Rothery's report, the Board of the North British Railway initially retained his services for rebuilding the bridge and for securing the necessary Parliamentary approval. It was, however, during the debate for a new Tay Bridge Bill that Bouch's reputation was finally seen as a hinderance to progressing the rebuild and he was asked to resign. His design for the Firth of Forth was sidelined, eventually in favour of Benjamin Baker's iconic masterpiece. He continued to work for a short time before being taken ill in July 1880. He died on 30th October, having outlived his Tay Bridge by only ten months.

5.3.   Tacoma Narrows – 1940

5.3.1   Context

Many books and papers have been written on the collapse of the Tacoma Narrows Bridge, but I have found none better than Richard Scott's book, *In the Wake of Tacoma* (Scott, 1956).

By virtue of the fact that it was caught on film, the collapse of the Tacoma Narrows suspension bridge over the Puget Sound in Washington State, USA, in November 1940, is one of the most widely known of all bridge collapses. It has become an essential part of any young engineer's education, not only in their understanding of structural engineering and aerodynamics but also in terms of the ethical responsibilities of their profession.

To fully appreciate the scale of the Tacoma collapse and its consequences, one needs to have some appreciation of the development of suspension bridge design in the early decades of the twentieth century. Although long-span suspension bridges were being built in other countries, there is no dispute that the lead was being taken by the USA, not only in terms of the bridges built but also the engineers who designed them.

It is also important to recognise where the Tacoma Narrows Bridge was placed in the contemporary hierarchy of suspension bridge spans; at the time of its opening, just four months before it collapsed, and with a main span of 853 m, Tacoma was the third-longest clear-span suspension bridge in the USA (and indeed the world) behind only the Golden Gate Bridge in San Francisco, completed in 1937, with a main span of 1320 m, and the George Washington Bridge over the Hudson River in New York City, completed in 1931, with a main span of 1100 m.

Of all the various structural forms of bridges, suspension bridges are, by intent, the most flexible. In terms of suspension bridge design, it was Leon Moisseiff who first introduced the use of the deflection theory as a way of determining the share of load between the bridge cables and deck, with particular relevance to stiffness. Moisseiff first used the deflection theory in his design of the Manhattan Bridge over the East River in New York, completed in 1909, and only the second crossing over that stretch of water after Roebling's Brooklyn Bridge opened in 1883. The theory itself had been developed by an Austrian academic, Josef Melan, in 1888 (Billington, 1983). The fact that Moisseiff was a European-trained engineer may have led to his enthusiasm for this approach to design. He was not alone, however, as one of the titans of American suspension bridge design, David Steinman, translated the theory into English in 1913 and used it himself in designs in the 1920s (Steinman, 1913).

A corollary of the theory was that the longer the span, the less the requirement for deck stiffness. In fact, Othmar Amman, another (if not *the* other) of the US suspension bridge titans, concluded that his design of the George Washington Bridge (1931) could, in theory, require no deck stiffness whatsoever. Another comparative factor to consider is the ratio between depth of stiffening member and span. For the original George Washington Bridge this was 1/350; for Golden Gate, 1/180. For Tacoma, this had grown to 1/400 and, interestingly, it was Leon Moisseiff himself who was also the designer of the Tacoma Narrows Bridge.

How to provide the requisite level of deck stiffness was also a key issue that had taxed the early designers. For long-span bridges, from Brooklyn Bridge onwards, the preferred structural form had been a truss, or rather a pair of trusses, below the main cables. As designs evolved over the first decades of the twentieth century, the depth of stiffening trusses reduced until the required stiffness could be achieved using plate girders, culminating with Tacoma and stiffening plate girders just 2.4 m deep.

Another factor with the Tacoma design was its width: designed only for two lanes of traffic, as opposed to its much wider contemporaries, meant that this aspect ratio was also relatively slender.

5.3.2   Aerodynamic behaviour and collapse

Since its opening in July 1940, the gentle vertical undulation of the bridge even under relatively low wind speeds had not only earned it the nickname of 'Galloping Gertie' but had also made it a tourist attraction for that very reason. Motorists crossing the bridge would see vehicles in front of them disappear into the trough of the wave and then reappear. While the general public may have been amused by its movement, concerns among engineers had begun even during the later stages of construction when a rhythmic vertical motion under only moderate wind speeds had first been identified. In fact, as a result, the University of Washington had been commissioned to develop a dynamic physical 1/100 scale model of the bridge. In parallel, similar motions had been identified in both the Deer Isle Bridge in Maine and the Bronx-Whitestone Bridge in New York City. Both of these bridges had been opened in 1939 and were designed respectively by Steinman and Amman, using the deflection theory and by utilising plate girders for deck stiffening.

Just a few weeks after Tacoma's opening, measures were implemented to monitor the displacements, with markers being attached to lamp columns and a movie camera positioned on the roof of the toll building. While detailed measurements of the amplitude and frequency were taken, unfortunately none of these had been correlated to wind speed or direction. Concern was growing before the anticipated arrival of stronger autumnal winds and additional holding-down cables were fitted, although clearly with limited success.

On the morning of 7th November, a hardly excessive storm had generated wind speeds of up to 68 km/h and it was at this point that the vertical undulations in the deck were augmented by a hitherto unseen torsional, rotational movement. As the morning progressed, the bridge was clearly in its death throes, and nothing could be done to control the destructive motion. At about 11 am, a section of deck about 183 m long peeled away and fell into the Narrows, leaving towers and remnant side spans as a sad reminder of what had been, just moments before, an elegant structure (Figure 5.4).

As Richard Scott points out in his unrivalled account, although no lives were lost (other than that of a dog in a car that had been left on the bridge), 'the biggest casualty was a generation of

Figure 5.4 The Tacoma Narrows suspension bridge collapse (courtesy of ICE Library)

engineering practice' (Scott, 1956, p. 53). The bridge had been designed to withstand wind speeds of up to 161 km/h and yet had spectacularly failed with speeds of less than half that figure. Leon Moisseiff was alleged to have stated that he was *'completely at a loss to explain the collapse'* (Scott, 1956, p. 53).

5.3.3   Investigation and consequences

To its credit, the US Public Works Administration (PWA) quickly assembled a 'Board of Engineers', latterly known as the Carmody Board after the head of the PWA, John Carmody. The Board consisted of three highly reputable engineers: two from the USA, including Amman, and one from Europe. The Board submitted its report just four months after the collapse and it was made public in May 1941 (Amman *et al.*, 1941). As well as providing a highly forensic analysis of the exact failure mechanisms, in terms of both cause and effect, the report also identified that there had been a number of similar collapses of narrow nineteenth-century suspension bridges in Europe which had not been widely reported at a professional level. Too late for Tacoma, but a salutary reminder that the lessons from history must be learnt.

The main findings of the report related to aerodynamics and the susceptibility of the bridge to torsional oscillations as well as vertical undulations. It recognised that the nearest analogy to a long, slender, low-stiffness bridge deck was an aircraft's aerofoil section wing and that aerodynamic effects over the latter could be applied to the former. While Tacoma had all of the above deck characteristics, it was far from the aerodynamic shape that would have been needed to avoid its failure. Here was the first recognition that aeronautical engineering knowledge of wing flutter and vortex shedding effects should also be applied to bridges.

As noted above, contemporary spans at Deer Isle and Bronx-Whitestone had also been experiencing wind-induced displacements at the time of the Tacoma collapse. At Deer Isle additional stay cables were added in an attempt to control movement but in a storm in December 1942, many of these snapped under violent oscillations and other damage occurred to the stiffening girders. As a result, yet more stays were added, producing what was described as a 'cat's cradle' of stays (Island Journal, 2014).

At Bronx-Whitestone, the interventions were far more visual. As well as additional stay cables from the towers to the main span, a new stiffening truss was added to each plate girder, *above* roadway level, and tuned mass dampers fitted below the deck (Frieder, 2019). Improved knowledge of aerodynamics and the need to relieve permanent load in the main cables allowed the trusses to be removed in 2003 in favour of glass reinforced plastic fairings affixed to the original deck stiffening plate girders to give an improved aerodynamic performance.

Suspension bridge design philosophy changed dramatically as a result of the Tacoma collapse. Deflection theory was no longer considered acceptable for long spans and deck stiffness, including torsional stiffness, was the order of the day. As well as the replacement Tacoma Narrows Bridge, which opened in 1950, it is perhaps Steinman's design of the Mackinac Bridge, opened in 1957 and linking the two peninsulas of Michigan state in the USA, which was the ultimate such example (Rubin, 1958).

An even greater step change in suspension bridge deck design began with the Severn Bridge in the UK, opened in 1966. This was the first to introduce a box-girder deck arrangement, closely akin to an aerofoil in cross section. Although there have been some notable suspension bridges built with stiffening trusses since that date, it is this form that has been preferred by designers across the world. All of those bridges can perhaps be considered legacies of the lessons learnt from Tacoma.

5.4. Point Pleasant – 1967

5.4.1 Context

As noted in the previous case study, the evolution of US long-span suspension bridges had taken a step change once Roebling's cold drawn wire and aerial spinning of main cables had been perfected with the Brooklyn Bridge's opening in 1883. While this new technique was used extensively in the design of new major crossings, those of smaller spans occasionally resorted to tried and tested elements and even slightly unusual designs. One such bridge was the Point Pleasant over the Ohio River connecting the states of West Virginia and Ohio.

5.4.2 The bridge

Also known as the Silver Bridge because of its aluminium paint scheme, the Point Pleasant Bridge was no small affair (Figure 5.5). Opened in 1928, its total length was 445 m with a main span of 213 m and side spans each of 116 m. It carried just two traffic lanes. Although the original design had apparently specified conventional wire cables for its suspension system, a cheaper alternative was selected using chains formed of steel eye-bars (West Virginia DOT, 2024).

It is not clear as to whether this was a complete redesign as the as-constructed bridge was effectively an unusual hybrid between a suspension bridge and a truss. The latter elements were above road level on each side of the bridge with the suspension chains forming the top chord at the lower levels of the catenary. The suspension chains were simply pairs of eye-bars, with pin connections providing the necessary flexibility in the structure. The stiffness needed for this arrangement could not have been easily achieved with a wire cable but was sufficiently provided by the eye-bar chain system. From the higher points of the suspension chain, hangers (or suspenders) connected the eye-bars to nodes on the truss. Another complicated detail was the fact that each tower was sitting on a hinged rocker bearing on top of the river piers. It was the chains, however, that were the critical elements as they had zero redundancy.

Figure 5.5 Point Pleasant Bridge (courtesy of Wikimedia Commons, public domain – this image is a work of a United States Department of Transportation employee, taken or made as part of that person's official duties. As a work of the US federal government, the image is in the public domain. URL: https://commons.wikimedia.org/wiki/File:Silver_Bridge,_1928.jpg)

5.4.3   15th December 1967

Almost 40 years after its opening, the bridge had been designated as a link to the interstate highway network and was now carrying 4000 vehicles per day, with about 20 per cent of these being heavy trucks. At about 5 pm on 15th December 1967, eyewitness statements reported a loud noise akin to a gunshot. This was the failure of an eye-bar on the north side of the Ohio side span. Once this element had failed, its neighbouring eye-bar slipped off the connecting pin and the catastrophic collapse mechanism was unstoppable. Domino like, from west to east, spans and towers collapsed and within less than a minute, the whole bridge had fallen. Thirty-one vehicles fell, either into the river or onto the Ohio bank. The death toll was 46, with another nine injured (Åkesson, 2008).

5.4.4   Investigations

As noted in the previous chapter, it was on 1st April 1967 that the US federal government created the National Transportation Safety Board (NTSB), consolidating all transportation agencies into a single body. Its current remit is to investigate all significant incidents in all modes of transport (NTSB, 2024).

The Point Pleasant collapse was its first major investigation into a bridge failure. The investigation was very thorough and included a detailed examination of the wreckage retrieved from the river. The final report (NTSB, 1970) is exemplary, detailing all of the various rigorous processes used in the investigation. Some sophisticated remodelling of critical elements and connections, as well as materials testing, also formed part of the analysis.

Attention focused on the failed eye-bar in the Ohio side span where a crack had propagated from the connecting pin. The cause of the crack was found to be a combination of stress corrosion and corrosion fatigue, partly attributable to the live load effects leading to cyclical rotations around the pin connection between chain links. The NTSB report acknowledged not only that such phenomena were unknown when the bridge was designed but also that it would have been virtually impossible to identify any early signs of such cracking during inspections.

Although the original 1927 design was found to be compliant with contemporary standards, there had clearly been significant increases in vehicle numbers and sizes since then, as well as axle loads. Interestingly, the factor of safety used throughout the original design calculations was just 1.5.

5.4.5   Conclusion

Although the Point Pleasant Bridge collapse was the subject of a robust and forensic investigation, and the subsequent reports made a valuable contribution to US bridge engineering, in the days before the internet, sharing the lessons internationally was generally only achievable through academic connections. That said, there were important lessons to be learnt, especially with regard to fatigue contributions to fracture mechanics. Perhaps even more importantly, in 1968, the collapse led to the US Congress directing the Secretary of Transportation to introduce National Bridge Inspection Standards (NBIS). Finally introduced in 1971, while the standard has developed over the interim, it is still in use today (FHWA, 2022).

To mark the 50th anniversary of the collapse, a documentary film was produced which brings home the individual personal tragedies, as well as some of the more technical matters, including the paucity of bridge inspections in the USA at that time (YouTube, 2017).

5.5.   Cleddau – 1970

5.5.1   Context

The concept of box-girder bridges in the UK goes back to the Victorian era and was first used by Robert Stephenson for his designs for the Conwy Bridge in north Wales and the Britannia Bridge over the Menai Strait between mainland Wales and the island of Anglesey. Opened in 1849 and 1850 respectively, both were part of the Chester to Holyhead Railway mentioned with regard to the Dee Bridge collapse in Section 5.1. above.

Fast-forward to the 1960s and the form was seen as a popular and economic option for multispan bridges, not only in the UK but also in Europe. A box has high torsional stiffness, however, and understanding the distribution of stresses arising from bending, shear and torsional effects is complex and demands more sophisticated analyses than might be used for more straightforward beam or truss designs. Here was a case of academic research into box girders and, even more so, the development of design codes lagging behind the national appetite for new bridges (Wood, 2018).

The first reported box-girder failure was in 1969. The fourth Danube bridge in Vienna, a continuous steel twin box of three spans and a total length of 412 m, was under construction using the free cantilever technique. On the evening of 6th November, the two cantilevers had just been joined at midspan, but full moment connection had yet to be achieved. A drop in temperature that night, together with consequential differential thermal effects, was instrumental in the failure of the bottom flanges which were overloaded in compression and buckled (Åkesson, 2008).

Other box-girder failures followed in the next few years, including Strom Bridge in Koblenz and at Zeulenroda, both in Germany, in 1971 and 1973 respectively. The former killed 13 and the latter, four. Both were under construction at the time of the collapse (Scheer, 2010).

In the UK, the subject of this case study, Cleddau Bridge, collapsed in 1970. In Australia in the same year, although for much more than just limitations of engineering understanding, there was also the collapse of the West Gate Bridge over the river Yarra in Melbourne (see Section 5.6 below).

Although the contemporary tabloid press and schoolboy humour used the phrase 'box-girder bridge' with much disdain, there followed a vital investigation into the causes of the collapses, leading to a step change in design philosophy both in terms of technical detail and procedural requirements.

5.5.2   Cleddau or Milford Haven Bridge

By 1960 the port of Milford Haven in southwest Wales was rapidly expanding, not least through the construction of an oil refinery and attendant infrastructure. The crossing of the River Cleddau at this time, however, could only be achieved by ferry. Later in the decade, a decision was made to build a two-lane bridge which was designed by joint consultants, Freeman Fox and Partners and Sir Alexander Gibb and Partners. Their design was a seven-span continuous single-cell steel box girder with an overall length of 820 m, with a main span of 214 m; this was to be the longest clear, unsupported span in the UK. The box was trapezoidal in section, and embraced the relatively new method of fabrication, in that it was welded construction throughout. Work began on site in September 1968.

Once the end spans had been completed using falsework, the erection process then used the free cantilever method, as had been used for the fourth Danube bridge.

5.5.3   2nd June 1970

By this date, two 77 m spans at the south end of the bridge had been completed and the third, also to be a 77 m span, had been cantilevered 61 m from the pier support. The final 150 t box section of this span, to meet the next intermediate support, was just about to be moved into position for erection, when the cantilever folded at the pier, and the free end fell some 30 m to the ground.

Four construction workers were killed and another five were injured. Fortunately, although this span of the bridge was being constructed over a public highway and residential properties, and there was a very close call with only some outbuildings being demolished, there were no casualties below the collapse (Figure 5.6).

Construction of the bridge was immediately halted and did not recommence until October 1972.

5.5.4   The investigation

The Cleddau collapse was the most significant in the UK for almost 100 years. The government (through the Department of the Environment, the Scottish Development Department and the Welsh Office) reacted by commissioning an inquiry. Chaired by Alec Merrison, Vice-Chancellor of Bristol University, its remit was to cover not only Cleddau but also the West Gate collapse which is covered in the case study in Section 5.6 below. It is interesting to note the exact brief (my italics): 'Inquiry into the Basis of *Design* and *Method of Erection* of Steel Box-Girder Bridges' (Merrison, 1973).

Therefore, as well as scrutinising what had caused the Cleddau collapse, the inquiry was to examine the appropriateness of codes and standards for design for both permanent works and all temporary conditions during construction. There was a degree of urgency here, as many other steel box-girder bridges were either in use, being designed or under construction in the UK. The Committee was able to instigate major programmes of research, through various universities and other agencies, both experimental and theoretical. The output was the Interim Design and Workmanship Rules (IDWR) which were promulgated through the (then) Department of Transport's standard BE 6/73 (Bridle and Sims, 2008). In fact, when the Committee issued an Interim Report in June 1971, it showed that, throughout the UK, there were 51 steel box-girder bridges in service and another

Figure 5.6 Cleddau Bridge collapse, Milford Haven (courtesy of ICE Library)

37 on the drawing board. The 51 each had to be assessed using the IDWR and 32 were found to be in need of strengthening.

The scale of the issues for the bridge community (clients, consultants and contractors) at that time cannot be underestimated. But, with public confidence in the profession badly shaken, not only was it a major problem but, in a sense, it was also one of our finest hours. With the Committee having the authority to marshal the country's best engineers and other specialists, it was a remarkably rapid response. The proceedings of a conference hosted by the Institution of Civil Engineers in 1973 gives a full account of the journey from failure to success (ICE, 1973).

As far as the Cleddau collapse itself was concerned, it was found that the diaphragm at the pier support had been under-designed in that insufficient stiffening had been detailed. It seems certain that the actual loads in the diaphragm were much higher than had been calculated, partly due to the additional transfer of shear force from the load paths through the inclined webs of the box. The issue was one only of design and no blame was attributed to the quality of the fabrication.

5.5.5   The legacy

As well as the success of the Inquiry in establishing the cause of the Cleddau collapse and the speed of response in producing the IDWR, another recommendation dealt with the need to improve and regulate the processes of technical approval and checking. There were four key aspects, all of which are still highly relevant today.

- An independent check is required of the engineer's permanent design.
- An independent check is required of the method of erection and the design of the temporary works adopted by the contractor.
- The clear allocation of responsibility is to be made between the engineer and the contractor.
- Provision by the engineer and the contractor of adequately qualified supervisory staff who must be present on site, with their tasks and functions clearly distinguished.

It is vital that these points must never be diluted. Indeed, it has been argued that they should apply to other engineering, or even building, structures especially those with increased complexity (Firth, 2020).

They also relate to our professional and ethical responsibilities, especially the final point with respect to engineering competence, some of which are addressed in Part 2.

5.6.   West Gate – 1970
5.6.1   Context

As noted in the previous section, the West Gate Bridge collapse, in Melbourne, Australia, came just four months after that of Cleddau and was also part of the remit for the Merrison Committee of Inquiry. Although often mentioned in the same breath, as previously hinted, the two were different in many respects. In engineering terms, the West Gate Bridge construction methodology was unique. Not only was it flawed in this respect from the outset, but it was also exacerbated by extremely difficult relationships, both contractual and, with regard to the influence of trade unions, industrial. Indeed, the last two recommendations from Merrison, summarised in Section 5.5.5 above, are probably much more applicable for West Gate.

For a highly objective and knowledgeable commentary on the West Gate collapse, the reader is encouraged to view the only one of the late Bill Harvey's series of 'Bridge of the Month' essays that was a video presentation, due to a broken finger restricting his typing ability, marking the 50th anniversary of the collapse (Harvey, 2020). As well as Merrison, the collapse was also the subject of a Royal Commission in the Australian State of Victoria. The Commission's report is not only essential and informative, as well as sobering, reading, but is also the definitive text on the matter on which much of the following has been based (Victoria, 1971). The Royal Commission report has also been summarised with very little loss of detail elsewhere (Bignell, 1977).

5.6.2   West Gate Bridge

As had been the case at Milford Haven, and countless other communities around the world, the decades of the 1950s and 1960s had seen a growing demand in Melbourne for an improved link across the Lower Yarra River, over and above the limited capacity provided by the Williamstown ferry. Negotiations with the federal government had commenced as early as 1958 but it was not until 1964 that a feasibility study into a tunnel or a high-level bridge was commissioned. The outcome was inconclusive.

By 1965, however, what was effectively a private company had been formed, the Lower Yarra Crossing Authority (LYCA), and an Act of Parliament of that name passed giving the company powers to develop a tolled crossing. The company had already commenced discussions with Maunsell and Partners of Melbourne, but by 1966, and to their credit, Maunsell acknowledged their own limitations (at that time) of big bridge design and recommended the use of the then world leader, Freeman Fox and Partners, albeit based in the UK.

A detailed design was developed for concrete approach spans and a central five-span steel box girder. Cable stays were to support a main span of 336 m. With a total of 28 spans and an overall length of almost 2600 m carrying ten lanes of traffic, this was to be an impressive bridge by any standards. Apart from the scale, and also bearing in mind that both were designed by the same firm, there were many similarities between West Gate and Cleddau, including the use of inclined webs in the box girder.

In 1968, tenders were invited for three separate contracts.

- Contract F – Foundations.
- Contract C – Concrete bridge works.
- Contract S – Steel bridges works.

Contracts F and C were awarded to a local Australian contractor, John Holland, while contract S was given to World Services Construction (WSC), a subsidiary of a Dutch company, Werkspoor Utrecht N V.

5.6.3   Construction

Work began in April 1969 and by September of that year, contract F was substantially complete. Contract C was making good progress and substantial completion was anticipated by March 1971. Contract S, however, had been delayed by disputes over steel quality specification, fabrication tolerances and labour issues. In order to prevent further delays, an agreement was reached whereby

WSC would continue to fabricate the box-girder elements, but John Holland would assume responsibility for their erection. This became Contract E.

Due to the facts that they were similar bridges and had been designed by the same company, the collapse in June 1970 of the Cleddau span had triggered a design review which had been undertaken by Maunsell in the UK. As a result, some limited additional stiffening of the box sections was designed and implemented, although not without an argument with the LYCA as to whether this was actually needed.

One significant difference in the detail of the two bridges, however, was in their fabrication. As noted, Cleddau was an all-welded construction whereas West Gate relied mainly on bolted connections. This was in part due to the chosen method of erection, in that for each span, the box girder was fabricated in two sections (each weighing about 1000 t), split longitudinally. The chosen method was for each to be jacked up to a height of about 50 m, either side of the piers, and slid into position before being spliced together. A construction process of this complexity would normally have been expected to have a contractual requirement for a trial erection. This could have been undertaken at ground level with minimal risk but, for whatever reason, this did not happen.

The first such jacking, between piers 14 and 15 on the south side of the river, began on 15th May 1970 and, although delayed by bad weather and industrial action, was eventually completed on 1st September. This had not been achieved without mishap, however, as each half of the box was effectively a channel section with little torsional stiffness. The warping distortion at midspan was significant, leading to a substantial buckling of the free edge of the top flanges. While attempting to join the longitudinal splice proved very difficult, it was eventually achieved by the use of some fairly brutal means, including (almost as a measure of last resort) the use of concrete kentledge in an attempt to counteract deflections.

5.6.4   15th October 1970

The equivalent span of the north side was between piers 10 and 11. The same construction process had been employed but, in an attempt to avoid the longitudinal splice problem, from the outset the contractor proposed the use of concrete kentledge atop each half-girder, to try to control the actual deflections. This was not overly successful, and indeed was a contributory factor to the collapse, as the buckle problem remained unresolved. Another attempt at reducing the buckle was proposed which involved removing bolts from a *transverse* splice near midspan. This was clearly an error of judgement because this effectively reduced the compressive capacity of the top flange. The unbolted half-span began to settle and rotate against its neighbour, which could not support the additional load for any length of time.

There followed some frantic attempts to correct the problem and refit the bolts. Senior engineers were called to climb to the top of the box sections, to review the situation and to try to redress the earlier error. By this time, however, the steel was yielding and the midspan vertical deflection increasing. Sadly, the opportunity to evacuate the span (and the site below) was not taken and, some 50 minutes after the bolts had been removed, the vertical deflection was so severe that the span pulled itself off pier 10 and fell onto the huts below where some of the workforce were having their break. As well as the men who were on the deck at this point, there was also a crane, other plant, and tanks of fuel causing a fire to add to the mayhem. Thirty-five men, either on or beneath the bridge, died (Figure 5.5).

5.6.5   The Royal Commission of Inquiry

As noted at the beginning of this case study, the investigation by the Royal Commission of Inquiry left no stone unturned and covered in great detail every issue and circumstance which led to the tragedy. It examined every piece of the presented evidence and ultimately produced a report in which hardly any company – client, consultants, contractors, subcontractors – or any individual could not be found to be to some extent at fault, either by their direct actions or by their failure to intervene.

Contractual interactions between parties had occasionally been dysfunctional, with poor working relationships and a consequent lack of cooperation and trust. There had also clearly been communication difficulties, not least with a designer on the other side of the globe with a nine-hour time difference. This was also in the days well before email and even fax machines, when telex messages and late night/early morning phone calls provided the only available options. Poor industrial relations, including demarcation disputes and strike action, undoubtedly led to delays, disruption and poor quality of workmanship, which in turn added to pressures on site staff as the programme slipped and costs mounted.

Listing a summary of specifics from the Royal Commission is unrealistic in this book but I can only restate the point made at the outset, that its report and that of the Merrison inquiry should be essential reading for all.

Work on the bridge resumed in 1972 and it was finally opened to traffic in 1978.

5.7.   Sunshine Skyway – 1980
5.7.1   Context

Although any bridge over navigable waterways could be vulnerable to collision from shipping, there may well be public misconceptions that either adequate protection measures should be in place or that it should be the vessel below which would be at greater risk. Neither of these assumptions are borne out by the fact that, since 1837, there have been well over 60 occasions when a ship has hit a bridge causing its complete or partial collapse, often causing multiple fatalities (Scheer, 2010).

This case study considers, as an example, the collapse of the I-275 Sunshine Skyway Bridge when hit by the *Summit Venture*, an empty phosphate cargo ship displacing almost 20 000 t, on 9th May 1980 (Griggs, 2022).

5.7.2   The Bridge
Also known as the Tampa Bay Bridge, the original Sunshine Skyway was part of a link spanning the bay on Florida's west coast, with a total length of over 22 km and a bridge length of some 7 km. The main structure over the shipping channel was a steel cantilever through truss with a clear main span of 482 m, with two side spans each of 76 m completing the central bridge. The remaining spans were lower viaducts of underslung truss construction. There were actually two parallel identical bridges: the first opened in 1954 and the second in 1971 to the west of the first bridge. Each carried two lanes of traffic and once the second bridge was opened the road was given the interstate highway designation I-275.

The main piers over the designated shipping channel had some modest protection in the form of fenders but there was nothing for the equally vulnerable side or approach spans. Between 1972 and earlier in 1980, there had been at least four incidents of ships hitting the fenders and another, also in 1980, when two ships collided near the bridge.

5.7.3   9th May 1980
It was standard practice for all vessels entering the Bay to be under the control of a pilot who was familiar with the tides and currents as well as the notoriously changeable local weather conditions.

On the morning of 9th May 1980, a pilot took charge of the *Summit Venture* at about 6:30 am to bring it into Tampa and headed towards the bridge. At this time there was only a light mist over the sea, but this soon turned to fog and rain, which was followed by a severe thunderstorm. Strong, squally winds at speeds of up to 110 km/h hit the ship which, because it was empty, was riding high on the water and it was pushed outside the shipping lane. Problems multiplied when the ship's radar failed and, despite the best endeavours of the pilot, at 7:33 am the ship struck the pier between the south side span and the first approach span of the southbound bridge. The pier collapsed almost immediately, followed by large sections of the main span cantilever truss, the side span and the first approach span. Some of the bridge ended up on the bow of the *Summit Venture* but there were no casualties on board. It was a different story on the bridge where vehicles were also driving in the bad weather conditions. Six cars, one pickup truck and a Greyhound bus fell into the bay, killing 35 people. Sadly, it appears that four cars and the bus drove off the remaining deck sometime after the collapse.

5.7.4   The investigation
The US National Transportation Safety Board investigated this event as a marine accident primarily, as any lessons to be learnt would apply mainly to that sector rather than to the management of the bridge. It published a report in April the following year (NTSB, 1981).

The report recognised the difficulties faced by the pilot who was judged to have taken all possible avoiding actions, other than abandoning the entry into the bay entirely, and was largely exonerated from blame. The report was also critical of the national weather service for not issuing timely warnings, and for the fact that there was no facility for warning traffic on the bridge that 'there was

danger ahead'. This relates to those vehicles which simply drove off the remaining bridge, as noted above. This recommendation seems slightly unrealistic, as any system of warning signs advising of a failed bridge would have been needed at various locations over the entire length of the bridge.

The report also criticised the lack of pier protection 'which could have absorbed some of the impact force or redirected the vessel'. In terms of lessons learnt by the wider international bridge community, it is this last point which served as a wake-up call, and many long-span bridge owners reviewed their risk analyses for ship collisions and retrofitted energy-absorbing protection to supports in the waterway, or directional 'dolphins' on the approaches to the navigable channel. As a result, in 1983 the Federal Highway Administration issued specific advice for vulnerable bridges (FHWA, 1983) making ship collision an essential design consideration for new long-span bridges.

5.7.5   Aftermath

Immediately after the accident, the older northbound bridge was converted back to a single lane in each direction. In 1987, a new Sunshine Skyway, a cable-stayed bridge with increased main span and higher clearance, was opened. It is protected from ship collision by substantial dolphins or concrete islands, a design concept now mirrored by most bridges around the world. The old bridges were demolished in 1990 (Figure 5.8).

This case study resonates with the similar collapse of the Francis Scott Key Bridge in Baltimore, USA, in March 2024. This is discussed in Chapter 8.

5.8.   Ynys-y-gwâs – 1985
5.8.1   Context

Unlike other case studies in this chapter, the Ynys-y-gwâs bridge collapse in 1985 in West Glamorgan, Wales, UK, claimed no lives. Another significant difference is that it catastrophically collapsed when there was no traffic on it. It is included here as it served as a wake-up call for the

UK bridge fraternity, as well as at an international level, to better understand what could lead to accelerated degradation of prestressed concrete bridges and what action was needed to prevent it (Woodward and Williams, 1988).

5.8.2   The bridge

Built in 1953, the bridge itself was rather modest, with a single clear span of 18.3 m, carrying just two lanes of traffic. It is in the Welsh village of Cwmafan spanning the river Afan. Although carrying only a minor road, it had been part of the highway authority's winter salting network. The deck was a segmental post-tensioned structure, comprising nine internal I-beams, each of which were formed of eight precast sections that were stressed together longitudinally. Transverse prestressing through the I-beams provided the lateral load distribution. There were also box-section edge beams supporting solid masonry parapets but separated from the main deck by service bays (Figure 5.9). Joints between the sections (about 25 mm wide) had been filled with cementitious mortar. Construction took place on falsework which was removed when all stressing operations had been completed. The deck had been waterproofed prior to surfacing at the time of construction (Figure 5.9).

Ownership of the bridge had changed in 1974, due to a reorganisation of local government, and unfortunately any records of its design and construction, as well as inspection history, were lost in that transition. Good records, however, existed from 1979 onwards, from which time the bridge had been inspected, albeit mostly superficially, no less than ten times. Inspections in the UK at that time did not translate into a bridge condition as such but generally recorded a schedule of any visible defects. Nothing was noted in inspection reports, however, for the deck soffit, such as rust staining, cracking, spalling or differential deflections, but that does not necessarily mean that none were present. The bridge had also been closely observed in 1981 when it had had to carry an abnormal load, but no concerns were recorded.

5.8.3   4th December 1985

The exact time of the collapse is unknown, but it seems to have been first noticed at about 7 am when a lone motorist turned to drive over the bridge and landed in the bottom of the V-shaped remnants of the deck. He was uninjured and his car was repaired, apparently for just £300.

All internal I-beams had failed at approximately midspan, although not all at precisely the same location. The edge beams and parapets remained intact. Fortunately, an early decision had been taken by the local authority bridge engineer, not to just simply remove the debris from the riverbed but rather to recover as many intact segments as possible. They also called on the services of the UK's Transport and Road Research Laboratory (TRRL) who had been commissioned in 1981 to undertake long-term research and development requirements in civil engineering (Civil Engineering Task Force, 1981). Analysis of the Ynys-y-gwâs beams contributed to that research.

5.8.4   The investigation

A block of the deck consisting of nine I-sections, three wide and three long, and the downstream edge beam were taken to TRRL's facility in Berkshire for an autopsy-like analysis. All other elements were also carefully moved to the local highway authority's depot where detailed records were taken. What was obvious was that all prestressing tendons were severely corroded at both longitudinal and transverse joint locations. While it was clear that corrosion was the primary cause of the collapse, the TRRL work took a more holistic view of all of the relatively short life of the bridge.

Figure 5.9 Cross section of the Ynys-y-gwâs bridge deck (courtesy of Woodward RJ and Williams FW, 1988)

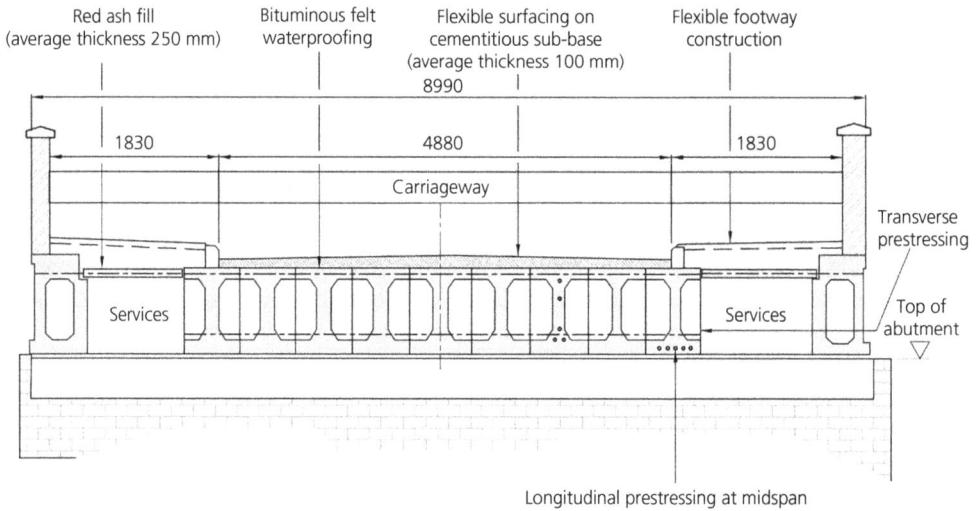

Firstly, it was clear that all the prestressing had been installed in accordance with the design and there was no suggestion that they had not been fully tensioned. Similarly, the anchorages were also in relatively good condition and the concrete in the beam sections was sound.

Attention soon focused on the joints between elements. Here it was found that the mortar packing had utilised high alumina cement, but the level of compaction was generally good. Where prestressing tendons crossed the joints, however, they should have passed through a metal sleeve which, although found in a few joints, in most locations a cardboard tube had been used instead. Neither material had offered much resistance against corrosion.

Within the beam sections, prestressing ducts, grout (both in terms of quality and volume within the ducts) and tendons were generally in good condition. Chloride contents were high throughout the tested sections.

A retrospective structural assessment was also undertaken which proved that bending capacity of the deck was greater than that under dead and live loading. There was just a marginal over-utililisation with respect to shear capacity, but this was thought not to have been a contributory factor to the collapse.

5.8.5 Conclusions

The biggest single factor that led to the accelerated corrosion of the prestressing tendons at both longitudinal and transverse joints was chloride penetration. The principal source of chlorides was from de-icing salts. The exact mechanism for the collapse was suggested as a sudden loss of capacity of one I-beam. As this failed it transferred its load to its neighbour which also failed and triggered a progressive collapse. Another contribution may have been temperature effects. There were no bearings as such, meaning that the deck was effectively restrained. As the temperature dropped and the deck tried to contract, the prestressing force would have reduced. Bearing in mind the season and time of day of the collapse, this may have been the proverbial straw that broke the camel's back.

5.8.6   Long-term legacy

Soon after the release of the TRRL findings, as well as input from the (then) Standing Committee on Structural Safety (SCOSS, 1987), there followed a comprehensive review of post-tensioned prestressed concrete bridges in the UK and an embargo on all such new bridge construction during the early 1990s. This still remains in place for segmental precast concrete bridges.

Another legacy, now part of national standards in the UK, is the need for formalised intrusive inspection of post-tensioned bridges, together with management planning to ensure that the risk of failure of such bridges remains minimal (National Highways, 2020).

5.9.   Conclusion

Although the very brief summaries of the collapses that have featured in these case studies have concentrated mostly on matters of civil and structural engineering, the most harrowing aspects of my research have been the personal tragedies of those affected by the disasters. This must be what we are all trying to avoid in our day-to-day bridge management.

While I have covered many of the causes, I have not attempted to make exact links to those that have been covered in Chapter 3 as this might be seen as being too prescriptive. The reader may draw their own conclusions here but, as the next chapter will show, there are still lessons to be learnt as we look at the twenty-first century.

REFERENCES

Åkesson B (2008) *Understanding Bridge Collapses*. Taylor and Francis, London, UK.

Amman OH, van Kármán T, Woodruff GB *et al.* (1941) *The Failure of the Tacoma Narrows Bridge.* HathiTrust Digital Library. https://babel.hathitrust.org/cgi/pt?id=mdp.39015021064251&seq=7 (accessed 03/01/2024).

Bignell V (1977) *Case Study 5, the West Gate Bridge Collapse. Catastrophic Failures.* The Open University Press, Milton Keynes, UK.

Billington D (1983) *The Tower and the Bridge.* Basic Books, New York, USA.

Bridle R and Sims F (2008) The effect of bridge failures on UK technical policy and practice. *Proceedings of the Institution of Civil Engineers, Engineering History and Heritage* **162**.

Civil Engineering Task Force (1981) *Long-Term Research and Development Requirements in Civil Engineering.* Science and Engineering Research Council, London, UK.

Cossons N and Trinder B (1979) *The Iron Bridge.* Moonraker Press, Bradford-on-Avon, UK.

FHWA (1983) Technical Advisory 5140.19. *Pier Protection and Waring Systems for Bridges Subject to Ship Collisions.* FHWA, Washington DC, USA. https://www.fhwa.dot.gov/engineering/hydraulics/policymemo/t514019.cfm (accessed 10/01/2024).

FHWA (2022) National Bridge Inspection Standards. https://www.fhwa.dot.gov/bridge/nbis.cfm (accessed 24/01/2023).

Firth I (2020) *Learning from History: Why the Box-Girder Bridge Failures Matter Today.* Institution of Structural Engineers, London, UK. https://www.istructe.org/resources/blog/learning-from-history-box-girder-bridges/ (accessed 19/01/2024).

Frieder D (2019) *The Magnificent Bridges of New York City.* Brilliant Publishing, Exton, USA.

Griggs F (2022) Tampa Bay (Sunshine Skyway) Bridge Disaster. *STRUCTURE magazine.* https://www.structuremag.org/?p=20417 (accessed 09/01/2024).

Harvey W (2020) *West Gate Collapse.* Bridge of the Month No 118. https://www.billharveyassociates.com/bom/118-westgate-collapse (accessed 29/01/2024).

ICE (1973) Steel box girder bridges. *Proceedings of the International Conference organised by the Institution of Civil Engineers* (February 1973), Thomas Telford, London, UK.

Island Journal (2014) The Year Steel and Cable Changed Deer Isle. https://www.islandjournal.com/history/year-steel-cable-changed-deer-isle/ (accessed 03/01/2024).

Lewis PR (2007) *Disaster on the Dee.* Tempus Publishing Ltd., Stroud, UK.

McKean C (2006) *Battle for the North.* Granta Books, London, UK.

Merrison AW (1973) *Report of the Committee of Inquiry into the Basis of Design and Method of Erection of Steel Box-Girder Bridges.* HMSO, London, UK. https://www.istructe.org/getattachment/a1301a4d-7acb-4f3e-9623-4f8d6e1c91c1/attachment.aspx (accessed 28/01/2024).

National Highways (2020) CS 465 Management of Post-Tensioned Concrete Bridges. National Highways, Birmingham, UK.

NTSB (1970) Highway Accident Report. Collapse of the US 35 Highway Bridge, Point Pleasant, West Virginia Dec. 15(1967) NTSB, Washington DC, USA. HAR7101.pdf (ntsb.gov) https://www.ntsb.gov/investigations/AccidentReports/Reports/HAR7101.pdf (accessed 23/01/2024).

NTSB (1981) *Marine Accident Report. Ramming of the Sunshine Skyway Bridge by the Liberian Bulk Carrier Summit Venture. Tampa Bay, Florida.* NTSB, Washington DC, USA. https://www.ntsb.gov/investigations/AccidentReports/Reports/MAR8103.pdf (accessed 09/01/2024).

NTSB (2024) *History of The National Transportation Safety Board.* https://www.ntsb.gov/about/history/Pages/default.aspx (accessed 01/02/2024).

Petroski H (1994) *Design Paradigms.* Cambridge University Press, Cambridge, UK.

Rubin LA (1958) *Mighty Mac.* Wayne State University Press, Detroit, USA.

Scheer J (2010) *Failed Bridges. Case Studies, Causes and Consequences.* Ernst and Sohn, Berlin, Germany.

SCOSS (1987) Standing Committee on Structural Safety. *Seventh Report of the Committee for the two years ending July 1987.* CROSS (UK), London, UK.

Scott R (1956) *In the Wake of Tacoma.* Republished in 2001 by the American Society of Civil Engineers, Reston, VA, USA.

Steinman DB (1913) *Translation of Josef Melan Theory of Arches and Suspension Bridges.* McGraw-Hill, New York, USA.

Swinfen D (2016) *The Fall of the Tay Bridge*, 2nd edn. Mercat Press, Edinburgh, Scotland.

Victoria (1971) *Report of the Royal Commission of Inquiry into the Failure of West Gate Bridge.* CH Rixon, Melbourne, Australia.

West Virginia DOT (2024) Silver Bridge. https://transportation.wv.gov/highways/bridge_facts/Modern-Bridges/Pages/Silver.aspx (accessed 23/01/2023).

Wood J (2018) Merrison revolutionised design after 1970 UK box girder collapses. *40th IABSE Symposium.* https://www.researchgate.net/publication/333905751_Merrison_Revolutionised_Design_after_1970_UK_Box_Girder_Collapses (accessed 29/01/2024).

Woodward RJ and Williams FW (1988) Collapse of Ynys-y-gwâs bridge. *Proceedings of the Institution of Civil Engineers, Part 1* **84**: 635–669.

YouTube (2017) 50th Anniversary of the Silver Bridge Collapse. https://www.youtube.com/watch?v=ErntxB0nLHc (accessed 23/01/2024).

Richard Fish
ISBN 978-1-83608-559-1
https://doi.org/10.1108/978-1-83608-556-020251006

Chapter 6
Twenty-first century case studies

Having discussed bridge failures and their consequences in earlier chapters covering the first two millennia AD, it might be hoped that in our present century such things would be behind us. As Table 4.1 in Chapter 4 sadly demonstrates, however, this is far from the case. As shown, and at the time of writing, at least 1252 people have been killed as a result of at least 71 fatal bridge collapses worldwide since 2000. Although records may not be as reliable before 2000, a comparison of the first 24 years of this century with the last 50 years of the twentieth century is given in Table 6.1, showing that, if there is trend in bridge collapses, it is going the wrong way.

What these figures also show is that the average number of fatalities per fatal collapse is reducing, possibly as a result of generally safer construction practices.

As with Chapter 5, the chosen case studies listed here are purely examples, taken in chronological order. There is also no significance of these examples over any others which could have been chosen, nor do any omissions signify any reflection that those were less important. Also as in Chapter 5, where there is a definitive reference, this is noted at the start of that case study.

Table 6.1 Comparable collapses worldwide and fatalities 1950–2024 (data courtesy of Wikimedia Commons, licensed under the Creative Commons Attribution-Share Alike 4.0 International Licence (https://en.wikipedia.org/wiki/List_of_bridge_failures)

Period	Number of fatal collapses	Number of fatalities attributable to the collapses
1950–1974	23	831
1975–1999	31	856
2000–2024	71	1252

6.1.   De la Concorde overpass – 2006

This collapse was another which benefited from being thoroughly investigated by a Royal Commission (Quebec, 2007), the report from which has been the substantive reference for this case study and is highly recommended reading for anyone eager to understand more about this collapse.

6.1.1   The bridge

The de la Concorde overpass was designed in the late 1960s as part of the expansion of the transport infrastructure around the city of Montreal in Quebec, Canada. It was considered to be innovative for its time, crossing a six-lane freeway, Autoroute 19, with a single span of 35.3 m. Supported by counterfort abutments and a 4 m in situ concrete cantilever slab from each, the principal structural element of the span was a simply supported suspended, or 'drop-in', section formed of prestressed concrete box beams placed side by side with a reinforced concrete top slab. There were two decks with a longitudinal joint between them, both supported from the two cantilevers by half-joints, with notional movement joints above. The overpass also carried a six-lane highway. One of the reasons for opting for the single-span solution was the fact that the autoroute beneath the bridge was on a horizontal curve and a central pier would have led to a significantly longer bridge to provide forward visibility for users of the freeway.

The main innovation in the bridge design was the maximisation of the single span using the precast suspended decks and the cantilever extensions from the faces of the abutments. The most critical elements, however, were the half-joints that facilitated that innovation and the reliance on the joints in the carriageway above them to prevent water and contaminants from reaching the bearing nibs.

6.1.2 Inspection and maintenance regime

In common with many structures of this age, inspections in the first few years of the bridge's life were, at best, rudimentary. Also familiar to other bridge owners managing bridges of that generation were issues of poor reporting, poor record keeping and loss of information during reorganisations of managing authorities. By 1992, when the bridge was barely 20 years old, an inspection revealed significant horizontal cracks extending from the internal corner of the east half-joint into the in situ cantilever slab, together with evidence of chloride deposits on the face of the concrete.

This inspection triggered a maintenance scheme which was completed in the spring of 1993. Works included the repair of the movement joints which in turn meant that areas of surfacing on the bridge decks had to be removed. Interestingly, the plant used to remove the blacktop was a swing shovel weighing in at over 40 t whereas the reference truck used in the contemporary design codes was 32 t. Whether or not this was a contributory factor in the collapse is unproven, but it certainly would have added a degree of overstress to some already ailing elements. Again, not unsurprisingly, when the surfacing was removed it became clear that the concrete near the joints was also defective and had to be replaced not only above but also well below the level of the reinforcement. The fact that the reinforcement was exposed to this extent also revealed some errors in the positioning of bars from the time of construction.

Despite the loss of concrete and the incorrectly placed reinforcement in the cantilever slabs at the half-joints, it seems that there was a disconnect between what was assumed to be purely a maintenance activity and the potential loss of load-carrying capacity of the structure. It is also unclear as to whether the specified waterproofing membrane was relaid on completion of the works.

Soon after these 1993 works, another reorganisation meant that inspection responsibilities transferred to another party. By 1995, inspection reports noted that the movement joints continued to leak and there was also a recommendation that the deck should be structurally assessed by no later than 1999. No such assessment was ever undertaken. In the 1997 inspection report, cracks were again noted on the visible faces behind the half-joints, although without any reference to number, locations, lengths nor widths. In 2002, the bridge's rating was reduced from 'good' to 'acceptable'. By 2004, however, without any intervention, it seems that the bridge had magically returned to a rating of 'good'. Despite this, and recognising his limitations, in the same year the inspector requested a special inspection. This was undertaken but what was missing was an accurate report of the 1993 works. That said, the outcome of the special inspection was that the visible cracks were 'not problematical' and that the structural capacity of the half-joint nibs 'did not appear a cause for concern'. Although there were two further inspections of the bridge after 2004, no further maintenance interventions were undertaken.

6.1.3   30th September 2006
At about 12.30 pm on Saturday 30th September 2006, the south deck of the overpass collapsed. Five people in two cars below were crushed to death. Another six driving over the bridge were injured.

Earlier that morning, reports had been received of concrete debris on the autoroute below the overpass. There had also been recent reports of a noticeable dip in the carriageway above from those driving over it. The existence of concrete debris had been reported to the highway authority and a Road Supervisor (with no bridge inspection experience) was dispatched to the scene, arriving below the bridge at about 11.45 am. There he noticed a 'chunk' of concrete (measuring 450 mm by 175 mm by 75 mm) and about 20 pieces 'about the size of golf balls'. Although the hole from which the larger piece of concrete had fallen was photographed, the debris was considered to be nothing out of the ordinary. The supervisor duly loaded it into the back of his truck and a request for an inspection of the bridge was submitted for the following Monday. The hole was on the outer face of the cantilever slab immediately behind the half-joint. The supervisor left the site at about 12 noon, only to receive a call just over 30 minutes later that more concrete had been reported on the autoroute. Five minutes after that, as he was returning to the site, he received another call that the bridge had collapsed. Once there, he confirmed that this was the case and, with a degree of understatement, advised that Autoroute 19 would have to be closed indefinitely.

As well as an immediate response from the emergency services, one of the first actions on the day of the collapse by the Montreal highway authority was to identify any similar bridges. Less than three hours after the collapse, its sister structure just 500 m north on Autoroute 19 was closed. Another 18 bridges were considered to be high risk and interim measures were taken before structural assessments could be completed.

6.1.4   The investigation
As noted at the beginning of this case study, as soon as Tuesday 3rd October, a Royal Commission was established. Three commissioners were appointed who undertook a thorough and forensic examination of all matters relating to the collapse, beginning with a visit to the scene on Thursday 5th October. Over the next fortnight, evidence was collected, samples taken and tested, and a rigorous inspection of the fracture planes completed.

The Commission not only studied the circumstances of the collapse but also carried out a detailed review of the bridge's design and construction. Here it found several dysfunctional organisations, a considerable degree of uncertainty as to where responsibilities lay, an overall lack of site supervision, and generally poor communication. The Commission was also critical of the bridge management regime after construction, including a lack of any as-built information, incomplete record keeping, a noncompliant inspection regime, and defects identified in inspections but not acted upon.

The Commission also studied the detailed design of the half-joint using the reinforcement in the nib as originally detailed compared with what had actually been fixed during construction. This work included some finite element modelling and full-scale tests undertaken at McGill University.

6.1.5   Conclusions
As with most bridge failures, the Commission was unable to find a solitary reason for the collapse. As well as a number of procedural recommendations, it listed both physical and human causes.

The report is at its most damming with respect to the latter. Looking back: '...*we now know that nearly 40 years ago, there was negligence on the construction site of the de la Concorde overpass, and lapses in the managing of the structure throughout its useful life.*' (Quebec, 2007, p. 7). And, looking forwards: '*To prevent such events, there need be consistent awareness and vigilance in relation to the demanding discipline of building or managing structures.*' (Quebec, 2007, p. 7).

Both statements reaffirm that we must all continue not only to learn but also apply the lessons learnt from events such as this.

6.2. I-35 W – 2007

As previously noted, there is probably no better national bridge collapse investigation and reporting organisation than the USA's NTSB. Similarly, there is no better source for this case study than its rigorous report from a highly forensic investigation (NTSB, 2008).

6.2.1 Context

Before jumping straight to the I-35W bridge, it is worth considering the context of the development of the US highway networks (LePatner, 2010). As early as 1916, at the dawn of the age of the internal combustion engine, the first legislation was enacted in which the federal government not only part funded strategic highway construction but also gave individual states responsibility for designing, building and maintaining roads and bridges which could eventually form part of a new pan-state network, albeit in compliance with national standards.

Road building grew rapidly in the 40 years that followed, with only relatively minor dips during the great depression and World War II. It was in 1944 that the outline of a strategic national network was first considered, and this became a reality with President Eisenhower's Federal-Aid Highway Act of 1956 which led to the formal creation of the interstate highway system. Optimistic estimates at the time believed that the whole network could be completed within 12 years.

There were two implicit problems with this rapid expansion. Firstly, capacity; while unskilled labour may have been plentiful, design consultants and construction companies were being stretched in an attempt to meet demand. Secondly, building new roads (and bridges) was always at the expense of maintaining existing, leading to a term often used by states in formal reports: 'deferred maintenance'. This was an acknowledgement that, although maintenance was needed, it could always be postponed to another time. This was to become a continuing malaise not just in the USA but in most countries in the developed world during this period.

6.2.2 The bridge

The I-35W design began in 1962 when the State of Minnesota commissioned consultants to develop a concept and then a final design for bridge 9340 (its number in the US national bridge inventory) over the Mississippi River in Minneapolis. With an overall length of 581 m, there were 11 approach spans of conventional construction and another three over the river. The concept of a twin truss bridge for those three main spans, with a total length of 324 m, was quickly accepted and early discussions centred on which grades of steel should be used in which elements, and how truss node connections should be made. The design and detailing were completed in June 1965. Construction began almost immediately and was completed in 1967 when the bridge was opened to traffic.

The choice of a steel truss was by no means innovative as the structural form had been popular across the USA, probably from the start of the railway expansion some 100 years before. Over the

72

40 years of the I-35W's life, traffic growth across the whole country was substantial. At the time of its opening, it was estimated that the bridge would carry up to 60 000 vehicles per day. By the time of its collapse this figure had reached 160 000.

Before delving more deeply into the relatively short life of the I-35W, it is pertinent to consider USA bridge condition classifications. There are various levels but probably the most worrying are those entitled 'structurally deficient', 'fracture critical' or 'functionally obsolete'. The last of these relates to a defect in a bridge's physical geometry, such as low headroom or narrow lanes. The second term means that there is no redundancy and a failure of one member, or connection, would lead to progressive failure or total collapse. Perhaps because of its somewhat dramatic title, this has recently been renamed as 'nonredundant steel tension member' (FHWA, 2022). The term 'structurally deficient' relates to a bridge's overall condition. In simple terms, the conditions of all the bridge's elements are translated into a sufficiency rating; 0 to 100. A score of 50 or below triggers the term 'structurally deficient'. A score of less than 50, and hence the structurally deficient tag, was attributed to I-35W in 2005.

Before 2005, however, there had been two major interventions: in 1977, the thickness of the concrete deck sitting on the trusses was increased from 150 mm to 200 mm and the number of lanes increased from six to eight. In 1998, concrete median (central reserve) and edge barriers were added, as well as improved deck drainage and additional bird guards, aimed at preventing birds from nesting within the structure and consequent damage to the paint system from their by-products.

The downward trend in the bridge's condition to 2005, almost certainly exacerbated by the additional dead loads added in 1977 and 1998, as well as the increase in vehicle numbers and weights, also meant that reaching 'structurally deficient' provided the state with an opportunity to bid for federal funding for maintenance and even replacement. It has been alleged that this level of deferred maintenance was a tactic adopted by State Departments of Transportation to secure additional funding (LePatner, 2010) although this has not been substantiated.

6.2.3   1st August 2007
Apart from some works being carried out on the bridge (seemingly insignificant, but not so), this was another normal summer's day. During the evening rush hour, however, at about 6.05 pm, there was a sudden and catastrophic failure within the main span. About 139 m of the bridge fell into the Mississippi, taking with it 111 vehicles. Thirteen people were killed and another 145 were injured. Figure 1.1 and Figure 6.1 below show different views of the collapse.

6.2.4   The investigation
Close on the heels of the emergency services, NTSB investigators arrived on site on 2nd August and remained there until 10th November 2007. As well as the formal report of the investigation, an interim safety recommendation was issued in January 2008 that all 'non-load-path-redundant steel truss bridges' should be assessed to ensure that steel stress levels were not in excess of those allowed by the federal standards. Significantly, this short recommendation made specific reference to gusset plates that formed connections at truss nodes.

The investigation was helped by the fact that the collapse had been captured on a nearby motion-activated CCTV camera, enabling 23 separate images to be examined over a time of 10 seconds. Close examination was conducted on the wreckage recovered from the river and a detailed assessment of the bridge's structural behaviour was carried out with a finite element analysis.

Figure 6.1 I-35W collapse (courtesy of Wikimedia Commons, public domain)

Figure 6.1 I-35W collapse (courtesy of Wikimedia Commons, public domain)

6.2.5 The cause of the collapse

In simple terms, the bridge collapsed because of a failure of a gusset plate. The main reason for this was identified as a design error compounded by the fact that the contemporary design review, or checking, procedures had been inadequate.

The report also reflects on what actions had led to the gusset plate failure. Firstly, the 1977 and 1998 interventions noted above had obviously added to the dead load effects. Secondly, however, it was the works in progress on the day of the collapse which had required four out of the eight lanes to be closed in order to replace the top 50 mm of the deck slab. A concrete pour had been scheduled for about 7 pm and construction plant and aggregate deliveries had arrived at the allocated location. It was a stockpile of aggregate which had been tipped at just the wrong position which was the proverbial straw that broke the camel's back. Among the report's recommendations is that the effects of any such work on a bridge should be checked by the owner and, as necessary, conditions imposed.

The gusset plate in question, however, had been found to have deformed well before the collapse. Examination of photographs from inspections suggested that this had occurred before 1999. The report also recommended that bridge inspections should include a greater emphasis on truss connections.

6.2.6 Conclusions

As shown above, the investigation into the I-35 W collapse seems to reflect several layers of complacency: firstly, with respect to the 1960s design and inadequate checking, perhaps also attributable to the amount of bridge design work at that time and the national lack of capacity mentioned earlier; secondly, regarding the vital part that inspections play and the need to ensure that inspectors are both competent and rigorous; and thirdly, when any works are being carried out on a bridge, there should be a detailed understanding of their impact on its structural capacity.

The overriding theme in all of the above is the need for competence at every level, including the bridge owner. It would be good to think that these lessons are being learnt; but are they?

6.3.   Viadotto Polcevera – 2018

It is unusual for a bridge to be known not just by its formal name but also by that of its designer. The Viadotto Polcevera (or Polcevera Viaduct) was one of these, and probably better known locally in Genoa, Italy, as the Ponte Morandi after the renowned Italian civil engineer, Riccardo Morandi (1902–1989). The bridge was, by any definition, a landmark structure, carrying the four-lane A10 motorway from the south of France towards Rome, some 45 m above the Polcevera river valley, just inland from the coast, over a total length of 1182 m. That a 210 m section of bridge should suddenly and catastrophically collapse in August 2018, in a wealthy Western economy, should have been unthinkable. But happen it did.

6.3.1   The bridge

As noted, the Ponte Morandi was an iconic bridge, considered to be not only part of Genoa's identity but also central to its economy. Riccardo Morandi was also a highly respected engineer, with an unparalleled reputation in his native Italy and for his legacy bridges across the world. He was a pioneer of post-war reinforced and prestressed concrete construction, with this crossing being his *pièce de résistance*, not only with respect to its main cable-stayed spans and pylons but also the trestle-like supports of the approach spans (Figure 6.2). It has been said that Morandi loved the 'poetry' of concrete (BBC, 2019).

Completed in 1967, the main spans were formed of concrete pylons with an integral deck and suspended, simply supported spans between them. The most unconventional aspect of these main

Figure 6.2 Polcevera Viaduct before the collapse (courtesy of Wikimedia Commons; this file is licensed under the Creative Commons Attribution-Share Alike 4.0 international licence (Share Alike 4.0 International - Creative Commons))

spans, however, concerned the cable stays. Not only was there just a single cluster of cables from pylon tops to the deck but also the fact that the 'stays' were encased in concrete, albeit prestressed, but only to a relatively low level of compression. Ironically, with regard to the latter, as well as his opinion that this solution improved the bridge's aesthetics, Morandi also saw this as a means of providing an enduring protection to the stay cables. Additionally, the single cluster of stays meant that there was no redundancy in the structure. Furthermore, and not uncommon in bridges of this generation, there had been no thought given to inspection and future maintenance.

Although the bridge was unusual, it was not unique. Morandi had first used the concept in his 1957 design of the General Rafael Urdaneta Bridge crossing Lake Maracaibo in Venezuela, opened in 1962 (Figure 6.3). With a total length of some 8.7 km and with a total of 135 spans, this was another major structure. It included five cable-stayed spans and six Polcevera-like pylons. Here, however, Morandi had opted for a more traditional stay-cable arrangement (albeit also single clusters), rather than encase them in his beloved concrete (Dupré, 1997). Although the bridge had itself suffered a partial collapse in 1964, this had been caused by an errant 36 000 t oil tanker which hit some approach spans and a 259 m length of deck had fallen into the water, taking with it four vehicles and seven lives (Visser, 2018).

Morandi reprised the concept once more in his design of the 282 m main span Wadi el Kuf Bridge in Libya, opened in 1972 (Figure 6.4). At that time, it was the longest cable-stayed span in the world and, until the Millau Viaduct in France was completed in 2004, with a height of some 160 m above the floor of the valley, it was also the world's highest bridge. Here, the stay cables were also encased in concrete, making it an almost exact copy of the Polcevera details. Problems identified in 2017, however, should have rung alarm bells in Genoa. Inspections had revealed some significant concerns over cracks found in the concrete sections, closely followed by recommendations for an immediate total closure. After some 'emergency maintenance', however, the bridge was judged to be sound enough to be able to remain open to light traffic but with security staff preventing trucks from crossing in groups (Libya Observer, 2017).

Figure 6.3 General Rafael Urdaneta Bridge, Venezuela (courtesy of Structurae)

Meanwhile, at Polcevera, the bridge had had more than its fair share of maintenance concerns. Unexpected levels of creep had been identified in the first few years after opening and this had been addressed by repairs to the concrete deck box beams. In 1979, Morandi himself had acknowledged that work would soon be needed to enhance the bridge's longer-term durability (Morandi, 1979).

Since its opening, there had also been dramatic changes to the governance and management of the bridge as, along with much of Italy's motorway network, this had effectively been subcontracted to the Società Autostrade. In 1992, its Technical Monitoring Group reported concerns at the top of pylon 11 which happened to have been the first to have been constructed (World Highways, 2019). In fact, the detailed assessment of exposed cables at the top of pylon 11 revealed significant corrosion and broken wires. It was the subject of a major maintenance intervention throughout the 1990s, which included new saddles and a total replacement of the stay system (Nuti *et al.*, 2020). The fact that similar interventions had been planned, but not implemented, for the other two main pylons would prove to have fatal consequences.

It was during the pylon 11 works, however, that the distinguished Polish engineer, Professor Janusz Rymsza of the Road and Bridge Research Institute in Warsaw, had the opportunity to visit the site. His main concern was not with the maintenance works but with the original design concept, recognising the inherent lack of redundancy in the structure. Rymsza even submitted a report, recommending the installation of additional stays below those of the original design (Research Outreach, 2022). Although Morandi himself had died four years earlier, Rymsza's suggestion was rejected, apparently on the basis of it being contrary to the designer's original aesthetic principles.

By 2011, the Società Autostrade had become Autostrade per l'Italia but was still the responsible body for managing the bridge. The company had reported in 2011 that the condition of the bridge was a cause of concern but suggested that its deterioration had solely been due to increased traffic weights and volume. Its report even suggested that a collapse within the next ten years was a possibility (Sood and Anwar, 2023). Another independent report by the Polytechnic University in Milan in 2017 identified specific issues with pylon 9 and its stays (La Stampa, 2018).

By 2018, as well as the reports of the problems with the Wadi el Kuf Bridge, it should have been clear that the chances of Polcevera reaching its intended design life, without significant maintenance interventions, were slim to say the least. As well as some poor detailing, probable 1960s workmanship issues and deterioration of the concrete through carbonation (not least due to its coastal location and pollutants from Genoa's industries), it was the zero redundancy of the concept which meant that any failure would be catastrophic.

6.3.2   14th August 2018

In the late morning of Tuesday 14th August 2018, Genoa was experiencing torrential rain and even occasional thunder and lightning. Poor visibility and surface water on the carriageway of the viaduct had slowed crossing traffic almost to a crawl.

At about 11.36 am, pylon 9 and its adjacent deck spans collapsed (Figure 6.5). It was later shown that the southwest stay failed explosively near its connection at the top of the pylon, leading to suggestions that the failure had been triggered by a lightning strike. As its load was transferred to the neighbouring northwest stay, that also failed almost simultaneously, taking the deck on that side of the pylon with it. The pylon itself was now asymmetrically loaded to such an extent that it started to rotate eastwards, progressively causing the east deck and then the pylon itself to collapse. The total time taken was later proved to be about 12 seconds (Calvi *et al.*, 2018). Some 30 vehicles were either on, or approaching, this section of the bridge at the time. In those vehicles, 43 people

Figure 6.5 Polcevera Viaduct collapse (courtesy of *New Civil Engineer*)

were killed and another 12 injured. Miraculously, there were no casualties in the various buildings below the bridge.

6.3.3   Investigation

In this instance, the investigation into the collapse was closely linked to legal proceedings. It has been said that the collection of evidence had begun within one hour of the collapse with the investigation being led by Genoa's Deputy Prosecutor, Paolo Davidio. Eighty individuals were summonsed for questioning, from the directors of Autostrade to relatively junior engineering technicians. Of these, 20 were suspected of involuntary or aggravated manslaughter and three were even placed under house arrest within weeks of the collapse. Davidio publicly expressed his view that this was a case of 'a criminal under-evaluation of risk' (BBC, 2019).

The mechanism of the collapse had been modelled based on CCTV footage which provided the basis for the sequence noted above (YouTube, 2018). There is little doubt as to what happened, and the evidence is also clear as to why it happened. The fact that the stays were encased in concrete made inspections of the critical elements impossible and the effectiveness of the concrete in protecting the cables from water ingress and corrosion was dubious to say the least. Corrosion had occurred and the progressive deterioration had meant that a failure was inevitable. It has also been postulated that low amplitude fatigue in the already corroded strands could also have been a contributory factor (Invernizzi et al., 2022).

It was to be almost four years before legal proceedings had reached the point at which a trial could take place, beginning on 7th July 2022 (BBC, 2022). By this time, the number of defendants had been reduced to 59 but with nearly all facing serious charges. At the time of writing, no conclusions have yet been reached.

6.3.4   Wider consequences

The Polcevera collapse triggered concerns with other bridges on the Italian strategic road network. Indeed, four other bridges had collapsed in Italy between 2014 and 2017, with two of these causing fatalities (Bazzucchi et al., 2018). In 2019 more than another 300 bridges were reported to be 'unsound'. The collapse also led other European nations to review the condition of their own strategic bridge stock – for example, in a government audit in France, 840 bridges were reported to be 'at risk of collapse' and, in Germany, the Federal Highway Research Institute reported that only 12.5 per cent of the country's bridges were in 'good condition' (Tolltrans, 2019).

Another point has been made concerning the perhaps unforeseen consequence of privatisation of a major infrastructure asset, in this case the Italian motorway network (Roberts, 2018). And Italy is by no means alone in taking this approach. Based on political dogma that the private sector is more efficient, and supposedly unencumbered by complex contractual arrangements, there is undoubtedly an extra layer of management and perhaps an obfuscation of the checks and balances needed to ensure accountability. The most important question must be: does privatisation make our bridges safer?

6.3.5   Conclusion

If there was ever going to be a bridge failure that should have been the wake-up call to first world countries, the 2018 disaster of the Viadotto Polcevera collapse should surely have been it. Given the structure's high profile, the tragic loss of life, and the calls in the technical press (Hansford, 2018 and Woof, 2018) for increased investment in bridge inspection and maintenance, there should surely have been a response and a commitment from Western governments. That, however, does not appear to have materialised. If this was indeed a wake-up call, then it seems that most politicians may have slept through it.

6.4. Nanfang'ao – 2019

Nanfang'ao (or occasionally written as Nanfangao) is a small, picturesque fishing port on the south side of the natural harbour of Su'ao, in the county of Yilan, on the east coast of the island of Taiwan. A relatively small community, Su'ao (including Nanfang'ao) has a population of only about 37 000. For such a relatively small place, it had a very big bridge which sadly collapsed in 2019. As noted elsewhere, Taiwan has its own Transportation Safety Board, and it is their report into this collapse which is the substantive reference for this case study (TTSB, 2020).

6.4.1 The bridge

Designed by the Chinese consultants, MAA, the bridge had been promoted by the Yilan County government to replace a lower-headroom bridge that had been restricting access to the port for larger fishing boats. It had been completed in 1998 and was owned and managed by the Taiwan International Ports Corporation. The bridge was a 140 m span steel tied arch, with a double fork arrangement at each end, supporting a two-lane roadway by means of 13 inclined cables from the arch to the central reserve separating the two carriageways. The deck consisted of a steel box girder with a composite reinforced concrete slab (Figure 6.6). It is worth noting that, by virtue of its location, the bridge was not only in a highly saline environment but also in a region of relatively high temperatures and humidity.

In the 20 or so years of its life, the bridge had become a tourist attraction in its own right, not least by the addition of artwork incorporated into the parapets, viewing areas outside of its pedestrian walkways and spiral staircases around each of the four supporting piers allowing access from deck level to the harboursides.

Figure 6.6 Nanfang'ao Bridge (courtesy of Wikimedia Commons, this file is licensed under the Creative Commons Attribution 2.0 Generic licence (2.0 Generic licence - Creative Commons)

6.4.2   1st October 2019

In the early hours of Tuesday 1st October, the Yilan region of Taiwan had been hit by a low-magnitude earthquake. By dawn that day, Su'ao was also still feeling the effects from the periphery of Typhoon Mitag which had crossed the island on the previous day, although local wind speeds had by no means been excessive.

At about 9.30 am, there was a solitary vehicle on the bridge: a fully laden fuel tanker heading east. As it neared the east abutment, the whole of the span collapsed. As it did so, the truck fell into the water below and caught fire. Unfortunately, also in the water below the bridge were three moored fishing boats and it was here that six fishermen were killed and about another 20 were injured. The truck driver survived, although he was badly injured. As with many twenty-first-century bridge failures, the moment of the collapse was captured on CCTV (YouTube, 2019).

6.4.3   Investigation

As noted, the Taiwan Transportation Safety Board conducted a detailed investigation into the collapse and its report was formally issued on 25th November 2020. The report concluded that neither the earthquake nor the typhoon had had any bearing on the bridge's demise, but it focused attention on the condition of the cables.

As the fuel truck passed the tenth cable (out of 13), it failed. The transfer of its load to its neighbours also led to their failure and to an almost instantaneous unzipping of the deck away from the arch. Without the support of the cables, the deck was drastically overloaded and collapsed.

Post-collapse inspections of the cables revealed significant corrosion, to the extent that the residual cross-sectional area was between only 22 and 27 per cent of that when the cables were installed. The main questions concerned how the cables had deteriorated to that extent in a little over 20 years and why had that deterioration in their condition not been identified. The former was clear: the cables had been encased in HDPE sleeves but, over time, the waterproof seals had failed, and the sleeves had become brittle and cracked. In its coastal environment, salt-laden moisture had been entering the sleeves for many years at a probably much greater rate than it had been able to escape. The investigation revealed that a significant number of wires in each cable had actually snapped well in advance of the collapse.

As for the lack of prior awareness of this problem, it seems that the bridge had only been the subject of cursory inspections over the course of its life. Although Taiwan's national highway authority had high inspection standards, these had not been applied by the bridge's owners nor had they employed competent staff to carry out inspections.

Another contributing factor related to the concrete deck slab thickness. Design drawings had specified a depth of 86 mm whereas the investigation revealed that the actual depth was 125 mm. The report concludes, however, that this issue, nor that of other minor detailing errors, had not been significant contributory factors to the failure.

Although the TTSB report does not specifically apportion blame, it has been reported that six individuals, from companies contracted to oversee construction and supervision, are facing charges of involuntary manslaughter and negligent injury (Taipei Times, 2022). Another allegation, of fraud, has been made in that documents are reported to have been forged and that unlicensed inspectors had been used for bridge inspections. At the time of writing, the outcomes from these prosecutions are not known.

6.4.4   Conclusion

As well as the obvious lack of a competent inspection and maintenance regime, due in part to an inexperienced client body, the bridge had clearly not been designed with maintenance in mind. Not only does it seem that there had never been a maintenance manual, but it also appears that it would have been impossible to have replaced a cable without closing the bridge to traffic, as well as also probably requiring significant temporary works. It could be argued, therefore, that the collapse was not so much due to the engineering mechanics but more to the issues associated with 'people' and 'process', not least bearing in mind the charges relating to negligence.

6.5.   Mexico City metro overpass – 2021

With a population of some 22.5 million people, greater Mexico City ranks as the seventh most populated city in the world. In terms of public transport, its citizens mostly rely on a rapid transit, or metro, network which carries around 4.5 million passengers every day. In the late evening of 3rd May, 2021, an overpass carrying Line 12 of the metro over Tláhuac Avenue collapsed as a train was crossing it. Twenty-five people on the train and a driver of a car below died and another 98 were injured.

6.5.1   Context

The Mexico City Metro system has 12 lines with about 226 km of track and 195 stations. The first line was opened in 1969 and the last (Line 12 or the Golden Line) in 2012. The Metro is a combination of underground and overhead tracks (Metro CDMX, 2024). Unusually, all but two of the lines use rolling stock with rubber-tyred wheels to reduce both noise and vibration; the latter aiming to reduce potential problems with foundations in the city's unstable geology. One of the two lines operating with steel wheels is Line 12.

Line 12, the longest line at over 24 km, was initially intended to be underground for its entire length, but land ownership and budgetary issues led to a decision to elevate almost half of its length, including the section in the borough of Tláhuac (Financiero, 2021). As well as some structural considerations, described below, another change during the construction was to switch to a different rolling stock with a slightly wider gauge. Another issue concerned the radii of some of the bends on elevated sections. Construction began in 2008 and was completed in 2012, a year later than programmed. It was fully operational in October 2012.

6.5.2   Elevated structure details

As might be expected, the design of the elevated elements for Line 12 followed a standardised approach for both design and construction. With simply supported spans of 30 m, the chosen deck was a reasonably conventional design: a pair of welded steel plate girders connected by K-frame bracing, acting compositely with a reinforced concrete slab which carried the twin tracks. With another recognised detail, composite action was assured by shear connectors welded to the top flanges of the beams. With spans of this size, it was almost inevitable that plate girders had to be spliced and this was achieved by in situ welding, at or about midspan.

This length of elevated structure also meant that there were elevated stations and, at some locations, this necessitated a localised widening. This was achieved by welding an additional beam to each of the plate girders to support the wider deck slabs which flared out to facilitate the widening.

A standardised design was also used for the single column substructure and pad foundation. Although column heights varied, it seems that the design had been based on a fixed maximum height. Occasionally, however, due to local issues with the vertical alignment, it seems that some columns were built to a greater height than had been used in the design.

6.5.3   Operations

Line 12 appears to have had more than its fair share of teething problems in its early years. Even before its official opening, monitoring during preoperational trials detected subsidence at several column locations. Staff working on these trials also noted excessive vibrations, especially in the Tláhuac borough area, with the vibration leading to a significant number of sleepers being damaged. The tight radii noted above also led to concerns of possible derailment and speed restrictions had to be introduced, as low as 5 km/h in some areas.

In March 2014, the elevated sections of Line 12 were closed for 18 months to enable track and structural repairs to be undertaken. As this was only 17 months after opening, an investigation was commissioned to determine the reasons for such poor performance. This was both an audit of documentation and some site testing and it revealed that errors had been made at all stages of Line 12's development: planning, design, construction and operation (Narayanan, 2021).

Shortly after the reopening, two trains collided, and 12 passengers were injured. In September 2017, Mexico City was hit by the magnitude-7.1 Puebla earthquake which led to 228 fatalities in the city alone. It also damaged much of Line 12's tracks and six of its stations. Although the line was reopened within six weeks, a number of columns were reported (apparently by members of the public) to have significant structural cracking. One such column was repaired in January 2018 by installing a collar around its base. Another defect was noted in which one of the main plate girders was found to have an excessive sagging deflection. The remedy in this case was to provide support to the beam with inclined props from the adjacent column base.

6.5.4   3rd May 2021

At about 10.22 pm an eastbound train on Line 12, about 75 per cent full, was approaching Olivos station in Tláhuac. Some 200 m from the station, the driver reported a jolt and a loss of power. The jolt had been caused by the collapse of the span that he had just crossed, taking with it the last two carriages of the train. The elevated track was about 5 m above the Tláhuac Avenue highway at this point and the carriages were precariously resting between the intermediate supports and the road below. As noted earlier, a total of 26 people died and 98 were injured to the extent that they had to be hospitalised.

6.5.5   The investigation

The investigation into the causes of the collapse was led by both federal and city lawyers and together they appointed Det Norske Veritas (DNV), the Norwegian risk management company, to undertake the investigation. Unfortunately, the immediate issues seemed to be more concerned with litigation and politics, even to the extent that there were arguments about whether a special commission should be established. Although DNV's initial brief was refreshingly wide ranging, covering every aspect of the design, construction, maintenance and operation, their findings were ordered to be split into three separate reports.

The first two reports focused on engineering issues and identified significant shortcomings with shear connectors both in terms of numbers and quality of welds, as well as more widespread issues with the quality of welds throughout the structure. They also found poor-quality concrete in the deck slabs, aggravated by poor workmanship and a lack of quality control and supervision (AP News,

2021). Unfortunately, when DNV's final report was issued in February 2022, the city government rejected its findings and even sued DNV over a perceived conflict of interest (Proceso, 2022). This meant that the report could not be released until the litigation processes had been completed.

A further litigious action was instigated against ten officials and supervisors, including the Line 12 Project Director, associated with the construction of the line with charges the equivalent of corporate manslaughter (Chicago Tribune, 2021). At the time of writing, it is understood that the trial has yet to be concluded.

6.5.6   Conclusion

This collapse seems to have encompassed almost every possible error that could have been made: an over-simplistic design concept, design errors, poor quality control and supervision, ineffective inspections, limited maintenance interventions, and the effects of nature. It also emphasises the importance of 'people' issues, which at best would seem to give an impression of a high degree of complacency and at worst, incompetence or negligence.

6.6.   Morbi – 2022

The collapse of the pedestrian suspension bridge in the city of Morbi in Gujarat in India in October 2022, with a death toll of 141, was the worst in terms of fatalities[1] since the Pamban rail bridge disaster, also in India, in 1964 which claimed 150 lives. This case study appears to be yet another in which complacency, incompetence and criminal negligence could have been significant contributory factors.

6.6.1   The bridge

Believed to have been built in or around 1880, the Jhulto Pul (literally the 'hanging bridge' in the local dialect) connected the districts of Mahaprabhuji and Saamkantha some 15 m over the Machchhu river in the city of Morbi. At the time of construction, it served to connect two palaces of the Morbi State royal family.

A traditional suspension bridge design with wire-rope main cables and hangers, although with no deck stiffening, the main span was 230 m but with a width between parapet fences of just 1.25 m. Unsurprisingly, this aspect ratio meant that the bridge was very lively and, allegedly, when first opened, there was a limit of no more than 15 people allowed on the bridge at any one time (Sharma, 2022). The bridge, described in the state's website as a 'technological marvel' (Pandey, 2022), became a tourist attraction in its own right, although often as one for thrill seekers who enjoyed the experience of its excitations.

In more recent times, the bridge was a tolled crossing, owned and operated by the Morbi city authorities. Day-to-day operation and maintenance had been contracted to the private sector, at least since 2008 when a local company, Oreva, was first commissioned. In March 2022 a new, 15-year, agreement was signed with Oreva for the overall management of the bridge which included staffing, ticketing, security and cleaning as well as maintenance. Presumably as part of this agreement, the bridge was closed in April 2022 for repairs and renovations, reported to have cost the equivalent of about £210 000. It is not clear, however, what those repairs and renovations entailed. Interestingly, Oreva's core business seems to have been clock manufacturing.

---

[1]There had been another disaster in 1983 in Ulyanovsk in the, then, USSR which claimed 177 lives. This was a ship collision with a rail bridge, but the bridge did not collapse, and the casualties were all on the ship.

Six months later, on 26th October 2022, the bridge was reopened with Oreva's Managing Director proclaiming that no further major work would be needed for another eight to ten years and (in a slightly odd remark) that, 'if people act responsibly', the renovation should last for 15 years.

6.6.2   30th October 2022

On the evening of Sunday 30th October 2022, in celebration of the festivals of Diwali and Chahath Puja (a local new year celebration), the bridge was packed with hundreds of people, including many families with children, very few of whom were using the bridge to actually try to cross the river. Some estimates put the figure on the bridge as high as 500, but a figure of about half that number is probably more realistic. In the age of smartphones, however, many were filming their celebrations, which included some who were deliberately intent on setting the bridge swaying.

At about 6.40 pm, the bridge failed catastrophically, almost certainly due to a sudden fracture of one of the main cables, throwing hundreds of people into the river below. Numbers of fatalities were finally reconciled at 141, including 55 children. Press photographs taken on the morning after the collapse appear to show only one of the main cables still in place but the deck is non-existent, over the full span of the bridge.

6.6.3   The investigation

Bearing in mind that the bridge had only just reopened after its refurbishment, attention was soon focused on the extent and quality of those works, on whether the bridge was actually in a fit state to be opened to the public and, bearing in mind the previous limits, to the number of people who had been allowed on the bridge.

With regard to the former, it appears that the deck timbers had been replaced, possibly with an aluminium deck, but no works had been undertaken on the suspension system (except possibly some repainting). The facts that the works had taken six months and at a spend of less than the equivalent of about £35 000 per month would seem to confirm this. Although it is improbable that the wire ropes of the cables and hangers dated from the original construction, eyewitnesses stated that the failed cable had been badly corroded (Scroll.in, 2022).

In terms of the timing of the opening, it seems likely that there was a desire to be ready in time for the aforementioned festivals, but questions of approval certification by the city authorities, the competence of those undertaking the works and the apparently unlimited sale of tickets for the crossing on the evening of the disaster clearly required answers. Fortunately, the Indian legal system was very quickly mobilised and less than a week after the disaster, nine people had been arrested and appeared in court. The nine ranged from Oreva managers and ticket clerks to others subcontracted by Oreva for the refurbishment and some security personnel stationed at the site (Tara, 2022).

A formal inquiry into the disaster was undertaken by a 'special investigation team', appointed by the Gujarat government, producing interim findings in December 2022 and a final report in October 2023 (The Wire, 2023).

In terms of engineering, the reports suggest that of the 49 wires forming the wire rope of the fractured main cable, 22 were so badly corroded that they may well have been broken before the failure. There are also suggestions that some of the hangers had been strengthened by the addition of stiffer bars welded to them. While this may have had a longer-term consequence, it is likely that this is immaterial in terms of the overall collapse.

The latter report is also very critical of the administrative processes, both within the Morbi city authorities and in the Oreva company, including failures to follow standard procurement procedures for the Oreva contract and the lack of any technical competence, either required by the contract or provided by Oreva.

6.6.4   Conclusion

The sheer number of casualties from this collapse still seems incredible and makes it such a personal tragedy for so many families. The reasons for this number probably lie in the totality of the failure and the fact that there was a 15 m drop to water level. It was also dark and there was very little in the way of emergency services readily available to deal with such an incident.

In this case, the engineering reasons are overshadowed by the degree of complacency and the almost total absence of any technical competence in the management of the bridge, and especially with regard to the carefree approach to the refurbishment and its reopening. The fact that senior managers are behind bars for the equivalent charge of corporate manslaughter should be a salutary reminder for those charged with managing bridges anywhere in the world.

6.7.   Conclusion

While the case studies in this chapter have reflected many engineering shortcomings, they have more in common with issues of 'people' and 'process'. If there are recurring themes, they are the need for competence at every level of the bridge management process, the absolute avoidance of complacency and the importance of responding to the evidence that is in plain sight; making the right decisions at the right time, no matter how unpopular those decisions might be to the public and politicians alike. As Table 6.1 shows, the number of fatal collapses so far in this century has exceeded those of the previous 50 years. If this is an indication of a trend, it needs to be addressed as a matter of some urgency.

REFERENCES

AP News (2021) Report blames poor welds for Mexico City subway collapse. https://apnews.com/article/caribbean-mexico-city-mexico-health-coronavirus-pandemic-7c94fb-9284796c96110a360961397805 (accessed 05/04/2024).

Bazzucchi F *et al.* (2018) Considerations over the Italian road bridge infrastructure safety after the Polcevera viaduct collapse: past errors and future perspectives. Frattura Integrità Strutturale – Academia.edu. https://www.academia.edu/37687501/Considerations_over_the_Italian_road_bridge_infrastructure_safety_after_the_Polcevera_viaduct_collapse_past_errors_and_future_perspectives?email_work_card=view-paper (accessed 07/05/2024).

BBC (2019) Assignment – Genoa's broken bridge – BBC Sounds. https://www.bbc.co.uk/sounds/play/w3csy5ct (accessed 20/03/2024).

BBC (2022) Genoa bridge: Hopes for new Italy as disaster trial opens – BBC News. https://www.bbc.com/news/world-europe-62076262 (accessed 24/03/2024).

Calvi G, Moratti M, O'Reilly G *et al.* (2018) Once upon a time in Italy: The tale of the Morandi Bridge. *Structural Engineering International.* https://doi.org/10.1080/10168664.2018.1558033 (accessed 22/03/2024).

Chicago Tribune (2021). India bridge collapse: 9 people arrested as Morbi death toll reaches 140. https://www.chicagotribune.com/espanol/sns-es-imputan-a-diez-exfuncionarios-por-caida-de-metro-de-mexico-20211206-2solptwusfck3c75bbxaj7xy2e-story.html (accessed 05/04/2024).

Dupré J (1997) *Bridges.* Black Dog and Leventhal Publishers, New York, USA.

FHWA (2022) *National Bridge Inspection Standards*. Federal Register, https://www.federalregister.gov/documents/2022/05/06/2022-09512/national-bridge-inspection-standards (accessed 08/03/2024).

Financiero E (2021) El desastre de la Línea 12 – El Financiero. https://www.elfinanciero.com.mx/opinion/dario-celis/2021/05/07/el-desastre-de-la-linea-12/ (accessed 04/04/2024).

Hansford M (2018) Maintenance is key to prevent another Polcevera. *New Civil Engineer*, (September 2018). EMap publications, London.

Invernizzi S, Montagnoli F and Carpinteru A (2021) Very high cycle fatigue study of the collapsed Polcevera Bridge, Italy. *Journal of Bridge Engineering*, ASCE, **27(1)**. https://ascelibrary.org/doi/full/10.1061/%28ASCE%29BE.1943-5592.0001807 (accessed 13/05/2024).

La Stampa (2018) Ignored university reports about Morandi bridge: 'Deformed pylons and oxidized cables'. https://www.lastampa.it/esteri/la-stampa-in-english/2018/08/17/news/ignored-university-reports-about-morandi-bridge-deformed-pylons-and-oxidized-cables-1.34039115/ (accessed 21/03/2024).

LePatner B (2010) *Too Big to Fall*. Foster Publishing/University Press of New England, New Hampshire, USA.

Libya Observer (2017) Authorities in east Libya close Wadi el Kuf Bridge for safety reasons. https://libyaobserver.ly/inbrief/authorities-east-libya-close-wadi-el-kuf-bridge-safety-reasons (accessed 20/03/2024).

Metro CDMX (2024) Metro CDMX. https://www.metro.cdmx.gob.mx/ (accessed 04/04/2024).

Morandi R (1979) The long-term behaviour of viaducts subjected to heavy traffic and situated in an aggressive environment: The viaduct on the Polcevera in Genoa. *Wayback Machine*. https://web.archive.org/web/20180818052325/https:/webapi.ingenio-web.it/immagini/file/byname?name=riccardo-morandi-durabilita-ponte-pp.pdf (accessed 21/03/2024).

Narayanan S (2021) *Collapse of the Mexico City Metro Overpass*. Indian Association of Structural Engineers. https://www.researchgate.net/publication/352868812_COLLAPSE_OF_THE_MEXICO_CITY_METRO_OVERPASS (accessed 04/04/2024).

NTSB (2008) *Collapse of I-35W Highway Bridge*. NTSB, Minneapolis, Minnesota, Aug. 1(2007). https://www.ntsb.gov/investigations/AccidentReports/Reports/HAR0803.pdf (accessed 07/03/2024).

Nuti C, Briseghella B, Chen A and Lavorato D (2020) Relevant outcomes from the history of Polcevera Viaduct in Genova, from design to nowadays (sic) failure. *Journal of Structural Health Monitoring*. https://www.academia.edu/93153493/Relevant_outcomes_from_the_history_of_Polcevera_Viaduct_in_Genova_from_design_to_nowadays_failure?email_work_card=view-paper (accessed 07/05/2024).

Pandey G (2022) Morbi bridge collapse: How India tourist spot became a bridge of death. *BBC News*. https://www.bbc.com/news/world-asia-india-63477292 (accessed 07/04/2024).

Proceso (2022) Sheinbaum demandará a DNV por peritaje "tendencioso y falso" del desplome de la Línea 12 [Sheinbaum to sue DNV for "tendentious and false" expert report on the collapse of Line 12]. https://www.proceso.com.mx/nacional/cdmx/2022/5/4/sheinbaum-demandara-dnv-por-peritaje-tendencioso-falso-del-desplome-de-la-linea-12-285412.html (accessed 05/04/2024).

Quebec (2007) *Rapport d'enquête sur l'effondrement d'une partie du viaduc de la Concorde Gouvernement du Québec*. Dalhousie University, Halifax, Nova Scotia. http://cip.management.dal.ca/publications/report_eng concorde overpass montreal.pdf (accessed 03/03/2024).

Research Outreach (2022) The collapse of the Polcevera Viaduct – An avoidable tragedy. https://researchoutreach.org/articles/collapse-polcevera-viaduct-avoidable-tragedy/ (accessed 21/03/2024).

Roberts H (2018) Italy's bridge disaster: An inquest into privatisation. *Financial Times*, London, UK. https://www.ft.com/content/874b7e4c-ac3f-11e8-94bd-cba20d67390c (accessed 05/04/2024).

Scroll.in (2022) Morbi Bridge cable rusted, tragedy could have been avoided if it was repaired, police tell court. https://scroll.in/latest/1036416/morbi-bridge-cable-rusted-tragedy-could-have-been-avoided-if-it-was-repaired-police-tell-court (accessed 07/04/2024).

Sharma S (2022) Morbi bridge collapse timeline: How Gujarat bridge collapse that killed 135 unfolded. *The Independent* (accessed 06/04/2024).

Sood S and Anwar N (2023) The Genoa bridge collapse: How and why? Beale & Co. https://beale-law.com/article/the-genoa-bridge-collapse-how-and-why/ (accessed 21/03/2024).

Taipei Times (2022) Six charged over deadly bridge collapse in Yilan. https://www.taipeitimes.com/News/taiwan/archives/2022/09/01/2003784548 (accessed 19/03/2024).

Tara R (2022) Main suspension cable failure likely cause of Morbi bridge disaster. *Engineering.com*. https://www.engineering.com/story/main-suspension-cable-failure-likely-cause-of-morbi-bridge-disaster (accessed 07/04/2024).

The Wire (2023) Morbi bridge collapse: SIT's final report blames administrative lapses, technical incompetence. https://thewire.in/government/administrative-lapses-and-technical-incompetence-behind-morbi-bridge-collapse-sit-says#:~:text=Morbi's%20municipality%20gave%20the%20repair,broken%20before%20the%20bridge%20failed (accessed 08/04/2024).

Tolltrans (2019) Situation Critical: Can tolling save the world's bridges? *Traffic Technology International* 2018 **6** (magonlinelibrary.com) (accessed 03/03/2024).

TTSB (2020) Final report released on Nanfangao sea-crossing bridge collapse. Taiwan Transportation Safety Board. https://www.magonlinelibrary.com/doi/full/10.12968/S1356-9252%2823%2940089-5 (accessed 09/02/2024).

Visser A (2018) The collision between 'Esso Maracaibo' & the bridge. https://www.aukevisser.nl/others/id1337.htm (accessed 21/03/2024).

Woof M (2018) Italy's horrific bridge collapse is a sign of a wider problem. *World Highways* (September 2018). Route One Publishing, New York, USA.

World Highways (2019) The lessons of the Genoa bridge collapse. *World Highways*. https://www.worldhighways.com/wh10/feature/lessons-genoa-bridge-collapse (accessed 21/03/2024).

YouTube (2018) Morandi bridge collapse footage comparison to simulation for validation. BCB (youtube.com). https://www.youtube.com/watch?v=NLaYrRXfe30 (accessed 23/03/2024).

YouTube (2019) The moment a bridge collapsed into a harbour in Taiwan (youtube.com). https://www.youtube.com/watch?v=ZC9_X2mEhqg (accessed 19/03/2024).

Part 2

Future-proofing

Richard Fish
ISBN 978-1-83608-559-1
https://doi.org/10.1108/978-1-83608-556-020251007
Emerald Publishing Limited: All rights reserved

Chapter 7
Understanding bridge condition

It may seem an obvious statement but if we need to know what we are looking for in terms of a well-managed bridge stock, and in taking a risk-aware approach to bridge management, then we also need to know how to get there. In order to do that, we need to understand where we are now. This chapter, therefore, will attempt to describe the present situation by considering the current condition and status of bridge stocks towards the end of the first quarter of the twenty-first century, primarily in the UK but also with relevance to other developed countries. Chapter 1 touched on the principles of sound bridge management, and this is a theme which will be developed further below.

7.1. Ageing bridge stocks

With regard to highway bridges, and based on the above opening premise, the place to start would seem to be with the status of generic bridge stocks in the Western world. Immediately after World War II, all European economies were on their knees, but by the 1950s they had recovered sufficiently enough to be able to consider not only repairing the wartime devastation to critical infrastructure but also to plan improvements and an expansion of their transport networks.

In the USA, with infrastructure physically undamaged by the war, the massive build-up of industrialisation resulting from the wartime economy was fuelling a demand for peacetime projects, including the facilitation of the new American dream: owning an automobile. This led to a demand for road space and, in turn, to the 1956 Federal-Aid Highway Act (National Archives, 1956), which committed the country to some 41 000 miles (or 66 000 km) of interstate highways.

In the UK, the notion of a national motorway network had first been conceived in 1946 and the initial element, the Preston bypass and now part of the M6, opened in 1958 (UKMA, 2024). As well as horizontal and vertical highway alignments being designed for higher speeds, another innovation was the absence of any at-grade junctions and the consequent adoption of the concept of grade-separated intersections, therefore leading to the need for more bridges. The motorways were also augmented by existing trunk roads to form a strategic road network which also required bypasses of bottlenecks, as much for the benefit of communities choked by ever-growing traffic volumes, as well as for those parties whose priority was journey time reliability. As well as these nationally strategic routes, local highway authorities had also identified cities, towns and villages in need of ring roads, relief roads or bypasses. Indeed, much of the UK's contemporary highway engineering capacity sat within County Surveyors' departments. In 1967, however, the government established regional Road Construction Units (RCU), sited largely with offices either within, or near, county council headquarters (Smith, 1985). RCUs became the drivers of the road-building programme for the following 15 years or so, until they were privatised in the early 1980s.

Any impression that bridges, whether in the UK, USA or any other country, only matter if they were built after World War II must be allayed at this point. Highway bridges from medieval times are still in use today across Europe and many from the Industrial Revolution onwards were being built with materials other than brick and stone. Cast and wrought iron were used in bridges at the beginning of the nineteenth century and were being replaced by steel, as the material of choice, towards the end of it. The early twentieth century saw a new composite material, steel-reinforced concrete, being developed and this was soon in common use for shorter-span bridges. Longer spans followed with the use of the method first patented in 1928 by the brilliant French engineer, Eugène Freyssinet (1879–1962), of prestressing concrete by means of tensioned wires or cables.

Of course, it must be remembered that only relatively few of those bridges built in the nineteenth century carried the public highway. The railway bridge stock had grown rapidly from the 1820s onwards, with the 1840s in the UK being described as the decade of 'railway mania'. With some spectacularly iconic exceptions, the vast majority of that century's railway bridges were of masonry construction, from small-span culverts to towering viaducts. The two things that the railway and highway bridge stocks have in common, however, is that they will both deteriorate, and they will both need maintenance and management. While each will need a slightly different skill set with different emphases, they will follow very similar principles (Cole and Fish, 2022; Ricketts, 2017).

7.2.   Concrete bridges – post-war details

The expansion of the motorway, trunk and local road networks in post-war UK, and similarly in most other countries, had to utilise the perceived best practice both in terms of structural forms, and in readily available materials. Reinforced and prestressed concrete were generally seen to be the most economic and the most durable solutions. The need for multispan overbridges, if only to span a four-lane dual carriageway or even a six-lane motorway, required either a continuous structure or a series of simply supported spans. Design of the former required a more rigorous analysis, initially relying on influence lines and hand calculation methods, usually on a metre-strip basis, before more sophisticated structural analysis software was readily available to consider, for example, the benefits of transverse distribution in a slab by using a grillage analysis. Other, more elaborate, forms such as concrete box girders soon followed, with the need to appreciate the effects of more complex concepts such as torsional stiffness and shear lag (Hambly, 1976).

The collective emphasis, therefore, was one of optimising the efficiency of a design and this was often best achieved by the use of some details, popular with designers in the 1960s and 1970s, which have since proved to be problematic (to say the least) when considering their legacy issues with regard to inspection and maintenance.

One such detail is the half-joint (Figure 7.1). Half-joints were introduced as a means of simplifying design by creating a simply supported suspended span in a multispan bridge. They were either formed of in situ concrete with the suspended span being cast against a suitable filler material placed against the previously cast section, or by utilising precast beams landed on half-joint nibs. Encased within the joint would be an elastomeric bearing strip, with steel dowels depending on whether it was fixed or free to move. A movement joint was formed in the flexible surfacing of the carriageway above. Not only are half-joints almost impossible to inspect (Desnerck et al., 2018) but the movement joints above them will inevitably leak, allowing chloride-laden water to be carried to the highly stressed and vulnerable lower nib. UK standards now require existing half-joints to be considered on a risk management basis (National Highways, 2020a). It should be noted that the cause of the collapse of the de la Concorde bridge in Canada in 2006, covered as a case study in Chapter 6, was due to a half-joint failure. In the UK, half-joints continue to give concern as they are highly unlikely to meet their design life requirements without significant intervention either through major maintenance or by the introduction of interim measures. A recent report into two

Figure 7.1 Typical half-joint (Author's own)

specific examples has even suggested that complete demolition was an option under consideration (Manning, 2024).

Another detail occasionally used in an arch but more often used at the lower end of intermediate substructure supports, especially inclined piers, was the concrete hinge. This was, again, popular with designers as it formed a theoretical pin joint rather than having to transfer bending effects from the support to the foundation and vice versa. The detail was first used by Freyssinet on his Pont Candelier rail bridge over the River Sambre in Montigny-le-Tilleul, France, opened in 1922 (Freyssinet, 1923). Although Freyssinet used the hinge to create a pin at the crown of his concrete arch, in more recent times this is a detail which is usually buried and, even if exposed, cannot be visually inspected with any degree of confidence to assess its structural integrity. Although Freyssinet was the original proponent of concrete hinges and initially gave his name to the detail, other variations on his theme have been used (Figure 7.2).

As concrete hinges grew in popularity and were introduced in other countries, subtle differences of design and construction were introduced (Schacht and Marx, 2015). Guidance in the UK, however, is available as to how the residual strength of a hinge can be assessed (National Highways, 2020b).

As well as detailing, early post-war concrete bridges occasionally suffered from a combination of any or all of poor workmanship, a lack of site supervision, and inadequate quality control of materials. The four 'C's needed to ensure durable concrete – cover, cement content, compaction and curing – were not always fully appreciated nor applied to all 1960s and 1970s bridge construction. Programme and contractual pressures also meant that where there had been a problem with, say, poorly compacted, honeycombed concrete exposed only when formwork was removed, the only pragmatic solution was to employ limited repair techniques, mostly by 'bagging in' (rubbing in a cement-rich grout, literally in a hessian bag), with a consequential loss of durability and potential longer-term corrosion and spalling issues that in due course would need significant intervention, although possibly not for some decades in the future.

Figure 7.2 Concrete hinge detail (Schacht and Marx, 2015)

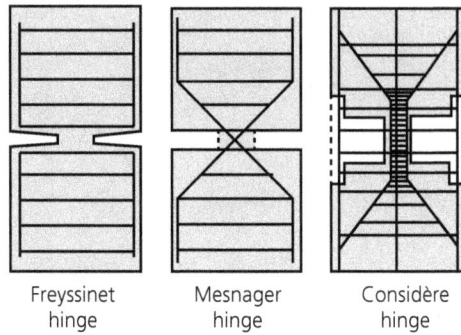

| Freyssinet hinge | Mesnager hinge | Considère hinge |

The weakest point in any reinforced concrete structure, however, will be the cracks which occur either as part of the design or the construction processes. Every engineer will know that the principle of reinforced concrete is that concrete is strong in compression but weak in tension. To overcome the latter, steel reinforcement is added to take the tensile stresses. However, in order for the reinforcement to be able to take those stresses, it has to elongate, and hence cracks will open. Similarly, other mechanisms such as shrinkage and plastic cracking will expose a weakness in the composite section (Concrete Society, 2010) which can result in accelerated corrosion over and above that which will occur purely through carbonation (Concrete Society, 2015).

Another shortcoming in concrete bridge detailing during this period was an oversight on how to effectively manage water. This applied not only to bridge deck waterproofing, which improved greatly from a 1950s hand-painted bitumen emulsion through to sheet systems and, most recently (and much more effective), spray-applied seamless membranes, but also to substructure drainage (an essential but often overlooked requirement to remove chloride-laden water from horizontal surfaces such as bearing shelves).

Lastly with respect to concrete bridges, it is those that employed Freyssinet's prestressing technique which have since proved to have given us some legacy problems. A distinction needs to be made here, however, between *pretensioning* and *post-tensioning*. The former is usually associated with precast beams of various forms (Trout, 2010) normally constructed in factory conditions with consequent improved quality control, in which the concrete is cast around stressed cables which are released when the concrete has reached the requisite strength. Post-tensioning is usually achieved by threading cables through ducts cast into a new concrete section (having attained adequate strength) which are then stressed to achieve a controlled compression and the ducts grouted to ensure that the cables are adequately protected. It is these which have caused concerns in the UK since the Ynys-y-gwâs bridge collapse described in Chapter 5 and are now required to be managed by taking a risk-based approach (National Highways, 2020c). Another post-tensioning technique, designed with maintenance in mind, is to use external strands usually inside concrete box girders. That said, such systems have also occasionally led to problems in recent years which have also required intervention (Stroscio *et al.*, 2024).

7.3. Steel bridges – post-war details

Although the majority of UK highway bridges built during this period were either reinforced or prestressed concrete, there were many which were either entirely steel or at least steel composite structures (such as steel beams acting compositely with a reinforced concrete deck slab). The former were relatively few and generally used only when significant spans were needed. The preferred form in these cases was the box girder, a bridge form not without issues, as noted in the case study on the 1970 Cleddau Bridge collapse in Chapter 5 and the subsequent Interim Design Working Rules (IDWR) (Merrison, 1973). The consequence of those rules meant that many box-girder designs then 'on the drawing board' had to be redesigned, and for those recently built or under construction, to be assessed and occasionally retrofitted with additional stiffening to ensure compliance.

A structural form little used in the UK but very popular in the USA (as noted in the I-35W collapse case study in Chapter 6) was the steel truss. The foremost exception to this generalisation are the main spans of the Boston Manor Viaduct, opened in 1965, carrying the M4 motorway just west of London. Unsurprisingly, politicians had to be reassured that this structure was not at risk immediately following the Minneapolis collapse in 2007. The same bridge, however, had to be severely restricted and emergency repairs undertaken in 2012 when fatigue cracks were discovered in some welds (Nicholls *et al.*, 2015).

Fatigue was also to be an issue on other major bridges, not least as it is an effect highly related to detailing, which again was not always fully appreciated in the immediate post-war decades. Similarly, by definition, it is also related to traffic volumes and heavy vehicle percentages, both of which have grown at a rate almost certainly more than had been anticipated. Several significant (and expensive) maintenance interventions have had to be undertaken as a consequence, such as that on the Gade Valley Viaduct opened in 1986 as part of the last link in London's M25 orbital motorway (Antoniou *et al.*, 2024).

Other problems with these longer-span bridges were not uncommon, usually relating to bearings and expansion joints. One example of the former was also found on the Boston Manor Viaduct in 2013 when an inspection revealed that it had almost rolled off its roller bearings (Arup, 2024) and another of cracked rollers on the M6 Thelwall Viaduct (Edwards *et al.*, 2009).

Perhaps the biggest issue associated with steel bridges concerned the need for protective coatings, otherwise known as paint systems. Repainting of steel bridges is not only essential to prevent corrosion and consequent reductions in section but it is also expensive: access, often requiring lane closures, environmental enclosure, scaffolding or mobile plant, meant that the material itself is only a tiny fraction of the total cost.

An alternative, popular from the 1970s onwards, was to use weathering steel (developed by USA Steel in the 1930s for use in ore wagons and registered under the brand name Cor-Ten, by which it is often still referred). This is a steel in which section sizes are marginally thicker to provide a thin sacrificial layer which slowly corrodes to form a red/purple/brown patina which becomes the protective layer (Dolling and Hudson, 2003). It was first used in the UK in 1967 for a footbridge at York University. The success or otherwise of weathering steel lies in the need for attention to detail – for example, ensuring that water cannot pond at web/flange/stiffener locations and that any deck drainage outlets are detailed to be located clear of the steel superstructure (SteelConstruction. info, 2018).

A salutary warning on the over-reliance on weathering steel as being a 'maintenance free' alternative, however, comes from the 2022 Fern Hollow Bridge collapse in Pittsburgh, USA (Figure 7.3). Opened in 1973, this bridge had been designed with weathering steel throughout, including for its

Figure 7.3 Fern Hollow Bridge collapse (public domain, courtesy of Wikimedia Commons/NTSB)

inclined piers which is where significant corrosion and section loss had occurred and the failure insti-gated, mainly because water had been able to pond against the steel faces. Fortunately, the collapse occurred early in the morning when there were only five vehicles on the bridge. Although there were no fatalities, ten people were injured (NTSB, 2024). An irony here was that the collapse occurred on the same day (28th January 2022) that US President, Joe Biden, was visiting Pittsburgh to promote his trillion-dollar infrastructure investment plan and later telling the media that the money would be found to repair some 43 000 bridges across the country (White House, 2022).

7.4. An ageing stock continues to age

As the above sections of this chapter have shown, our post-war bridge stocks have some inher-ent legacy problems which will continue to challenge bridge managers. The inevitable additional problem, however, is that those bridges will continue to age and, as well as the problems we are aware of, there may be others which will fall into the category of 'unknown unknowns'. In other words, problems may yet present themselves over which we have no current understanding but are most likely to be in terms of limited material longevity, such as those mentioned in Chapter 3 under 'unexpected material deficiency'.

In terms of ever-increasing age, however, and on the assumption that carbonation depth is propor-tional to the age of the concrete, it is inevitable that depths of carbonation in concrete bridges will increase and eventually reach the steel reinforcement. When this happens, the steel will corrode and expand, leading to spalling and the ongoing cycle of further deterioration. Concrete repairs will be needed to manage the rate of that deterioration. Similarly, although paint systems now have a much longer life than those in the past, steel bridges will still need painting to ensure that they will reach their design life.

Figure 7.4 shows an indicative representation of a typical bridge stock (in this case based on UK numbers but equally applicable elsewhere) showing the approximate numbers of bridges built every five years for the last 100 years or so. Projecting the curve of that profile forward 50 years

Figure 7.4 Projected future maintenance profile (courtesy of National Highways)

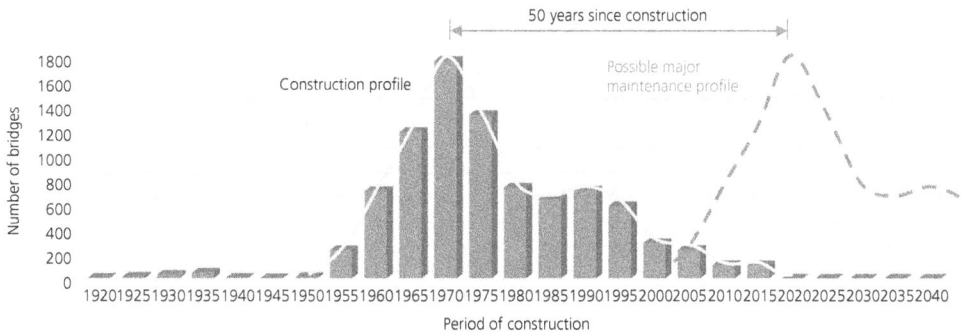

would suggest that spending on maintenance will have to match the equivalent investment in real terms that went into bridge construction in the post-war decades. While there is an argument that this is an over-simplification, it is highly unlikely that this will represent an underestimation of the likely demand for maintenance investment. As well as budgetary issues, the question of capacity within the sector will continue to be a challenge for all involved: bridge clients, managers, consultants, contractors and specialists alike.

7.5. Hidden and vulnerable details

As noted, legacies from the post-war bridge-building boom include many details, such as half-joints and concrete hinges as discussed above, which, thankfully, are probably not to be found in this century's bridge designs. Among the many challenges facing bridge managers, however, is to appreciate which details require additional attention and to understand the range of defects that may be developing within them.

As we have seen, many such details are hidden within the structure and their condition, therefore, cannot be assessed purely by visual inspection. Although this chapter has concentrated primarily on concrete and steel bridges, other older structures, such as masonry arches, timber bridges and those supported by cables, are by no means immune to having hidden details, the condition of which cannot easily be determined.

While masonry arch bridges may give the impression of solidity, many will have had historical interventions to correct issues such as spandrel movement through the installation of tie bars and pattress plates. The condition of the tie, however, even if galvanised, once encased in the fill, can never be assessed without some form of non-destructive testing. There are many other potential defects, however, which are too numerous to mention here but are well worth some additional reading (Harvey and Harvey, 2016).

With regard to timber bridges, the variability of the material itself can often mean that a defect is completely hidden within the timber section. In August 2022, the 10-year-old Tretten Bridge in Øyer, Norway, a truss with a slightly odd combination of weathering steel and glulam timber, collapsed with just one car and one truck crossing it at the time (Figure 7.5). Although the truck was understood to be close to the maximum live load for which the bridge had been designed, the official investigation concluded that failure had occurred in one of the timber diagonals which was

said to have had only about half of its theoretical capacity due to a 'material anomaly' (bridgeweb, 2022).

While some details will be fundamental to the structural integrity of a bridge, others may, at first sight, seem to be of lesser importance, such as connections for parapets or overhead signs. Although these will hopefully not be a contributory factor to a catastrophic collapse, they could easily lead to significant consequences, including serious injuries or even fatalities. These are termed 'safety critical fixings' and in the UK some research was instigated, and a report published (Stacy *et al.*, 2019), following the 2012 collapse of roof panels in the Sasago tunnel in Japan which claimed nine lives[1] (BBC, 2012). A similar situation in the UK, which, thankfully, was identified in good time, is covered as a case study in 7.13.3 below.

Returning to those hidden details which relate to the majority of the UK's post-war bridge stock, a seminal piece of work was undertaken in 2017 which should be an important part of any bridge manager's reference and reading matter (Collins *et al.*, 2017).

7.6. Overview of the UK bridge stock

Concerns over the condition of the UK's strategic bridge stock, specifically with respect to concrete bridges, are not new. In 1989, recognising that the majority of its stock was concrete, the then Department of Transport (before the days of the Highways Agency, Highways England or National Highways) commissioned a report on the performance of concrete motorway and trunk

---

[1] NB This collapse of panels within the tunnel has *not* been included in the statistics given in Tables 4.1 and 6.1.

road bridges. Commonly known as the Maunsell Report, this was the result of a survey of some 200 bridges in England, from Cornwall in the south to Northumberland in the north (Wallbank, 1989).

The report identifies issues as might be expected, both from a visual point of view but more importantly relating to those concerns noted above with respect to cover,[2] concrete quality, curing and the depth of carbonation,[3] plus external effects such as those arising from chlorides, sulphates and Alkali-Silica Reaction. Among its recommendations were many which relate to bridge inspection practice and the importance of record keeping, which will be discussed in 7.11 below. Others focus on how to improve the longer-term durability of new bridges, notably how to improve the process of water management: from rain landing on the carriageway to how it needs to be rapidly moved off the bridge, away from contact with concrete surfaces, and into effective drainage systems.

7.7.   Back to the USA

As has been noted in Chapter 4, the USA not only has the best investigatory body for reporting failures but also (as Table 4.1 has shown) has had proportionately more than its fair share of collapses in the twenty-first century.

An advantage (or even perhaps a disadvantage?) that the USA has over just about every other country is that it has a national bridge database: the NBI (National Bridge Inventory). As of 2023, across the 50 states and five protectorates or territories, there was a grand total of 621 581 bridges (FHWA, 2023). The construction date profile of that stock is again mostly post-war and, apart from the specific numbers, with a very similar age profile to that of the UK as shown in Figure 7.4. The accuracy of the 2023 NBI number, however, cannot be guaranteed because, for example, the argument of what span constitutes a bridge will also apply here, as well as those many other minor bridges on local roads which have not been identified as part of an individual state's contribution to the NBI. That said, there will be at least that 621 581 figure.

A 2023 report by ARTBA (the American Road and Transportation Builders Association) looked into the details of the NBI and made comparisons with 2022 (ARTBA, 2023). The findings are shown in Table 7.1.

The first point to note from these figures is that over 900 bridges have been added to the inventory, presumably from new construction. Secondly, although the number of bridges in good condition has dropped, those in a fair condition has gone up and that amounts to almost 50% of the stock. Other highlights from the report indicate that over 222 000 bridges need repair, including 76 600 that should be replaced. A cost of $319 billion is estimated for such a replacement programme. Perhaps more worrying is the fact that there are 167 million vehicle crossings of structurally deficient bridges every day. Another staggering statistic is that those 222 000 bridges, if placed end to end, would stretch almost 10 000 km.

Such figures reveal the scale of the problems facing bridge managers, not just in the USA but across the developed world. But there should be no insouciance about tackling these issues. It is imperative that every possible effort is put into managing bridge stocks as effectively as possible with a full awareness of both risks and our ethical responsibilities as professional engineers. These topics are covered in Chapters 8 and 9 respectively.

---

[2] The Maunsell Report noted cover ranging from zero (exposed bars) to 130 mm.
[3] The Maunsell Report noted maximum carbonation depths of 20 mm.

Table 7.1 USA National Bridge Inventory analysis (courtesy of ARTBA)

	Number of bridges	
Rating	2022	2023
Good	276 282	275 093
Fair	301 355	304 026
Poor (Structurally Deficient)	42 951	42 391
Totals	620 588	621 510*

*The small difference in the total number of bridges recorded by ARTBA and in the NBI reflects the different dates that the NBI was accessed.

7.8. Bridge management

The above sections have revealed the scale of the problems facing developed countries with respect to their ageing bridge stocks. The important point, however, is to understand what can be done about it and to take the appropriate action; and that is bridge management.

Many of the case studies covered in earlier chapters have shown that a contributory factor to the causes of a collapse proved to be ineffectual bridge management. Defining bridge management is not easy because it requires a holistic approach to the subject, as well as a specific focus on individual aspects, needed to provide a sound basis for decision making. It is not, however, merely a piece of process, just a set of procedures that need to be followed to achieve the right outcomes. It also requires a high degree of vigilance, closely followed by significant amounts of engineering judgement, pragmatism, and even scepticism in the form of a questioning mind, challenging the assumptions that may be being made from the data and information provided.

Bridge management also requires a degree of professional competence throughout; every member of a team must be trained in the tasks they have been asked to perform as individuals and must act diligently within their skill set as well as understanding their limitations. They must also be suitably encouraged to ensure that they maintain their appropriate level of skill and, when ready, be prepared to take on additional responsibility through continuing professional development.

Of course, these points about bridge management process and people represent the perfect situation which must be recognised as being highly unlikely ever to be achieved. It is important, however, not to have an underlying acceptance of the status quo. That would be the start of the slippery slope towards complacency; and complacency will be a root cause that will lead to a failure or collapse.

The need to raise the profile of bridge management in the UK was recognised in the early years of this century and, in 2005, a code of practice was published under the auspices of the UK Bridges Board (UKBB, 2005). The Bridges Board was one of four boards with specific interest in various highway asset management sectors and each reported to the UK Roads Liaison Group (UKRLG[4]). The code of practice was revised in 2016 but is now only available online (UKRLG, 2016).

---

[4] UKRLG is now the UK Roads *Leadership* Group.

Despite saying that bridge management should not just be a piece of process, there are indeed the rudiments of a process and the key elements that constitute that process are discussed below.

7.9.   Inspections

The essential requirement for a rigorous bridge inspection regime cannot be over emphasised. The main immediate post-war guidance on bridge design and construction in the UK, actually drafted *during* World War II (HMSO, 1945), had this requirement which says it all.

> *'The regular inspection of bridges is a matter of great importance, since the early detection of trouble and the prompt application of the appropriate remedial measures may well obviate costly repairs which may be needed if defects are allowed to develop too far.'* (HMSO, 1945, p. 30)

It is interesting to note that the above guidance is not prescriptive but gives an implicit trust in the competence of professional engineers to use their judgement as to how best to achieve those outcomes. Unsurprisingly, guidance for inspections in the UK has since developed in the interim to become more standardised (National Highways, 2020d) but at the same time giving useful advice for inspectors (TSO, 2007). This is the Bridge Inspection Manual, which at the time of writing, is being updated with publication anticipated in early 2025.

General Inspections of UK highway bridges should take place every two years, interspersed with Principal Inspections every six years. Similar timings apply to rail bridges, although these are referred to respectively as Visual and Detailed Examinations. The additional requirements for a Principal Inspection, over and above those of a General Inspection, are that every part of the structure should be accessed within touching distance and that they should be either undertaken, or led, by a Chartered Civil or Structural Engineer. Because a Principal Inspection can require a considerable amount of access equipment, and hence expense, the current Design Manual for Roads and Bridges (DMRB) standard allows a relaxation of the period between Principal Inspections to up to 12 years but based on an evaluation of the risk in doing so.

Recognising that inspections provide the bulk of the data and information upon which all bridge management decisions are made, it is essential that they are both accurate and comprehensive. The prerequisite in achieving these objectives is that inspections must be undertaken by competent people. The question, however, is how can such competence be proven?

Traditionally, bridge owners have employed inspectors 'from the tools'; skilled artisans who may have been masons, bricklayers, carpenters and so on. They would have been 'trained' by a more senior inspector who had probably also followed that same path. Such grandfather rights would mean habits – good and bad – would be passed down to successive inspectors. While there is every chance that this would produce competent inspectors, in order to *prove* that would require some form of independent assessment. This was a challenge considered in the UK by the UK Bridges Board and the Bridge Owners Forum in the 2010s and led to the creation of the Bridge Inspector Certification Scheme (LANTRA, 2016). The scheme requires inspectors to submit their experience before being interviewed by independent assessors. Once an inspector is certified, they are deemed to have met the nationally recognised standard. Although some other schemes are in place on a regional basis in the UK, few reflect the truly independent assessment of competence which is essential if inspection reports are to be trusted to inform decisions on whether, or when, a maintenance intervention is going to be needed.

7.10. Assessments

As well as having an appreciation of a bridge's condition, to have the confidence that it will perform as expected, it is equally important to know its structural, load-carrying capacity. The most recent update of the DMRB introduced a new standard, CS 451 Structural Review and Assessment of Highway Structures (National Highways, 2020e). This not only formalises the concept of a structural review but also requires that it should be undertaken at least at every other Principal Inspection.

A review must consider issues such as any change in loading standards and changes to highway geometry, as well as whether there has been any excessive or unexpected deterioration of condition. The outcome of a review is binary; either a recommendation that a structural assessment should be carried or that an assessment is unnecessary at this time.

Assessments are also subject to standards, notably CS 454 Assessment of Highway Bridges and Structures (National Highways, 2020f) although additional guidance is readily available in terms of principles (Shave and Bennetts, 2022) and for specific bridge types such as masonry arches (Gilbert *et al.*, 2022), concrete (CBDG, 2007) and steel (Slade, 2023).

7.11. Record keeping

In the context of an ageing bridge stock, individual bridge records are a fundamental aspect of sound bridge management. In an ideal world, every bridge would have its own maintenance manual, containing everything from original design calculations, as-built drawings, and details of every inspection and maintenance intervention since its construction. Sadly, however, for the majority of post-war bridges, this is highly unlikely to be the case. Often driven by such trivial demands as a lack of storage space, many such paper records will have been disposed of on the premise that they will never be needed again. The technology of the 1970s was to transfer drawings onto microfiche but usually, by that time, the paper drawings had not only been well thumbed but also starting to fade and becoming illegible (at least usually the important bits!).

Another issue, and one already noted in the de la Concorde overpass collapse case study, is that records are often lost in the transfer of maintenance responsibility from one managing agent to another. Although this should have been inexcusable, it happened all too frequently.

Even if there is an absence of earlier documents, inspection records should be retained and examined in order to identify any trends in deterioration. This emphasises the needs not only for competence in the inspection team but also for inspections to be objective and not subjective – for example, a significant crack should have both width and length recorded so that, when next inspected (even if not by the same inspector) a comparison can be made to see if that defect, and/or the condition of the bridge as a whole, is at a steady state or worsening. This point was also exemplified in the de la Concorde case study when subsequent inspections showed the bridge's condition went from 'good' down to 'acceptable' and then back to 'good' again without any intervention.

7.12. Maintenance interventions

In the perfect world, maintenance interventions would be both proactive and reactive. The former would include examples such as planned deck rewaterproofing (obviously coinciding with planned and coordinated carriageway resurfacing), replacement of bearings and expansion joints at the appropriate time, and the painting of steel elements before excessive corrosion occurs. Reactive maintenance will usually be driven by inspection data or from third-party damage such as from a vehicle hitting a parapet.

Unfortunately, and as noted in Chapter 1, and specifically Figure 1.4, this vision of utopia is certainly not, and is far removed from, the real world. If this was the case, then the bridge manager's life would be far less stressful.

Those readers who are, or have been, a bridge manager will know that the job is far from straightforward. In fact, more often than not, it can be the opposite. The challenges can be both relentless and intense, with often difficult and urgent decisions to be made against the backdrop of unsympathetic senior managers and political masters, and ever-declining budgets.

Returning to the topic of record keeping, in order to deal with such pressures, sound advice is to make extensive notes of all decisions taken, including the constraints within which they are having to be made. Such records can be useful evidence when lobbying for enhanced maintenance budgets.

Similarly, with regard to competence, and as noted above, it must be recognised that this is a characteristic which must apply not only to inspectors, but also to everyone in the bridge management team, including bridge managers themselves.

7.13. Close-call case studies
Underpinning this chapter on the need to understand and appreciate the implications of a bridge (or stock) condition has been the need to recognise precursor events and to react and respond to them, before a serious failure comes to pass. The following brief case studies are examples of some of those close-calls or near misses, the possible consequences and the actions taken to overcome them.

7.13.1 Churchill Way Viaducts, Liverpool, UK
Conceived in the early 1960s as part of an urban masterplan for Liverpool city centre, the twin Churchill Way Viaducts were multispan post-tensioned concrete flyovers. The northern structure was 239 m long, while its southern neighbour had a total length of 285 m. The deck section was a multicell concrete box supported on circular columns integral with the decks but with bearings at footing level. They were opened to traffic in 1970 and were given a Concrete Society award in 1971 (roads.org, 2020).

Although subjected to the contemporary inspection regime for their early life, it was only in 2018 that significant structural issues were identified. As an immediate interim measure, a 7.5 t weight restriction was imposed, pending a more detailed inspection and intrusive investigation which took place in 2019. That revealed significant concerns arising from the original construction with honeycombed concrete, incorrectly placed reinforcement and some movement of internal permanent formwork that had occurred during the original deck concrete pours. The inspection also found evidence of significant water penetration into the box cells and some external spalling, although the latter had been noted in inspections dating from 2009. The overall outcome of the 2019 inspection was dramatic; that both viaducts should be closed to both vehicles and pedestrians with immediate effect (Sholli, 2019).

Although a feasibility study was commissioned into methods for strengthening the viaducts, the decision was taken that they should be demolished along with a remodelling of local transportation links in the city centre. Demolition took place in 2019.

The root cause of the structures' demise was clearly down to poor workmanship, and presumably ineffective site supervision, during construction. Arguably, this is possibly exacerbated by an over-complex design. Perhaps more concerning, however, is the fact that inspections for the first 40 years of the viaducts' lives had not picked up the progressive deterioration. Fortunately, however, problems were identified, and action taken (albeit somewhat extreme), before any serious or even catastrophic failure.

7.13.2 Hernando de Soto Bridge, Memphis, USA

Opened in 1973, the Hernando de Soto Bridge (Figure 7.6) is a twin tied arch bridge carrying the I-40 highway over the Mississippi River between West Memphis in Arkansas and Memphis, Tennessee (Watts, 2021). The main spans are each 274 m, and the total length is 2875 m. The bridge was the subject of a major maintenance intervention in the form of a seismic retrofit which was completed in 2015.

In May 2021, during a routine inspection, a major fracture in a lower chord of the deck stiffening truss was discovered (Figure 7.7). The inspector who found this was so concerned that he called 911 and, with significant consequent disruption, the bridge was immediately closed not only to highway use but also to river traffic passing beneath (Action News 5, 2021a).

While at first sight, this may seem a good news story of action being taken as soon as a problem has been identified, it later transpired that the fractured chord had been photographed as early as 2016 by a father and son kayaking on the Mississippi (Action News 5, 2021b). A review of archived footage taken from a UAV (unmanned aerial vehicle) 'inspection' in 2019 by the Arkansas Department of Transportation also clearly showed the defect. The conclusion was also clear: that the FHWA inspection requirements (FHWA, 2022) of two-yearly, 'hands on' inspections had not been followed. As well as disciplinary action taken against the inspection team, a detailed investigation report revealed that the fracture was mainly attributable to poor welds dating from the original construction (Arkansas DoT, 2021). Following a detailed structural assessment, the river was reopened to marine traffic after three days. Steelwork repairs were quickly implemented, and the bridge opened to road traffic in August 2021 (CNN, 2021).

Figure 7.6 Hernando de Soto Bridge (courtesy of Wikimedia Commons. This file is licensed under the Creative Commons Attribution-Share Alike 4.0 International licence (Share Alike 4.0 international licence – Creative Commons))

Figure 7.7 Crack in lower chord of the Hernando de Soto Bridge (photo courtesy of Wiss, Janney, Elstner Associates)

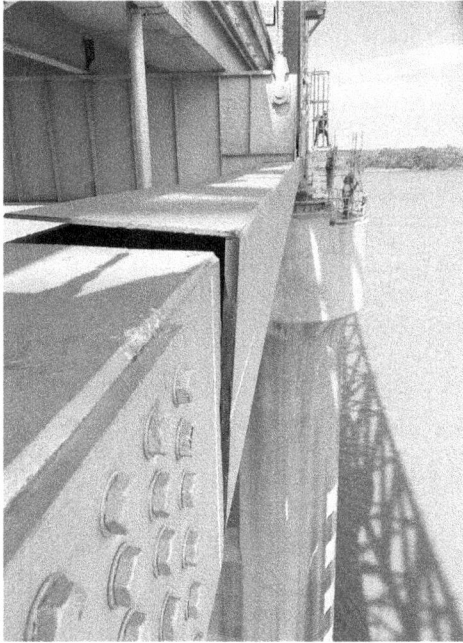

7.13.3   Balcombe Tunnel, West Sussex, UK

The Balcombe rail tunnel was opened in 1841 and is 1034 m long. The twin tracks within it carry train services between London and Brighton on the south coast. The tunnel was driven through a clay hillside but, despite being brick lined (at least six rings thick), it has suffered from water ingress throughout its life. In order to alleviate the not insignificant operational problems that this caused, five steel water catchment tray structures (four being 20 m long and the fifth, 72 m) were installed in the late 1990s above both tracks to catch the water and redirect it to the tunnel's drainage system. The trays were supported by transverse beams which in turn were held in place by resin fixings drilled into the tunnel lining.

At 5.24 am on 23rd September 2011, the crew of an engineering maintenance train spotted that one end of one of the transverse beams was 'not connected'. The route was immediately closed. An emergency inspection found that in fact three such beams had become detached at one end, with significant deflections. It also revealed that a total of 18 fixings were missing and another five were loose. To the credit of all concerned, the line was reopened after emergency repairs had been completed just 22 hours after the problem had first been spotted.

The incident was the subject of a rigorous investigation (RAIB, 2013) that concluded that the resin fixing material was incompatible with the brickwork lining, an issue exacerbated by the water ingress problem and the fact that, in some cases, an insufficient quantity of material had been used. The investigation also found evidence that some of the fixings had previously been found on the tunnel floor during maintenance work, but no action had been taken.

Although, this was primarily a case of failure of safety-critical fixings, the investigation report also noted the problems that asset managers have in being able to inspect and maintain such structures when the operational timetables mean that there is insufficient time to do so. Had one of the trays fallen onto the line, the consequences of the incident would have been huge, almost certainly worse than the Sasago tunnel failure noted in Section 7.5 above. It is interesting to note that the initial precursor events, that fixing studs had previously been found within the tunnel, were not reported at the time, demonstrating the need for both vigilance and an inquisitive mind in all parts of the workforce.

7.13.4   Hammersmith flyover, London, UK

This flyover in west London carries the A4 dual carriageway over the Hammersmith gyratory system and other local roads. It is 620 m long with a total of 16 spans. It was opened to traffic in 1961. It was a radical design for the time not only for the use of prestressed, post-tensioned concrete but also in the fact that all the deck elements were precast (Rawlinson and Stott, 1963).

The collapse in 1985 of Ynys-y-gwâs bridge, as noted in Chapter 5, had triggered a UK-wide requirement for post-tensioned special inspections (PTSI). Hammersmith flyover was the subject of a PTSI in the early 1990s which found significant corrosion of tendons and some fractured wires, caused by an ingress of chloride-laden water.

The PTSI was repeated in 2009 although with a much more intrusive investigation which found that the overall condition had deteriorated to the extent that almost half of the precast segments tested were in poor condition (Pearson-Kirk, 2022). A year later an acoustic emission monitoring system was installed to detect individual wire breaks. In the first three years, over 700 breaks were recorded. A deterioration model was developed which sought to extrapolate the worsening condition to the point that intervention was needed. The model predicted that reduced load-carrying capacity in the most critical span would be reached in November 2011, just nine months before the 2012 London Olympics, when every arterial road in the capital had to be unrestricted.

An innovative short-term strengthening scheme was devised and implemented, using a system of external post-tensioning. This was later supplemented by a system of additional prestressing which was designed to completely remove any reliance on the original system.

Hammersmith flyover is an exemplar in bridge management: from an awareness in the 1990s that condition was deteriorating, to testing the rate of deterioration in 2009, to then modelling the likely intervention point; at the same time, acquiring data on wire breaks to validate the theory and lastly to intervene in good time.

7.14.   Conclusion

It should be clear from all of the above that understanding the condition, either of a bridge stock or an individual bridge, is fundamental to sound bridge management. Condition is dependent on data collection from inspections. While there is a bridge management process, it would be a mistake to think that it is just that. Most importantly, there is the need for competence, capability and capacity in a bridge management team, and, above all, no room whatsoever for complacency.

REFERENCES

Action News 5 (2021a) I-40 bridge closed indefinitely after crack discovered in structure. https://www.actionnews5.com/2021/05/11/emergency-roadwork-shutdowns-i-bridge/ (accessed 08/05/2024).

Action News 5 (2021b) Photos show I-40 bridge damage in 2016. https://www.actionnews5.com/2021/05/18/photos-show-i-bridge-damage/ (accessed 08/05/2024).

Antoniou K, Bonnett J, Robinson P *et al.* (2024) The fatigue enhancement of Gade valley viaduct box girders due to distortional effects. *Proceedings of the Institution of Civil Engineers – Bridge Engineering* **177(1)**: 32–42.

Arkansas DoT (Department of Transport) (2021) 1-40 Hernando deSoto [sic] Bridge, Fracture Investigation.pdf. https://www.ardot.gov/wp-content/uploads/2021/11/Fracture-Investigation-I-40-MS-Rvr-Bridge.pdf (accessed 09/05/2024).

ARTBA (2023) ARTBA 2023 Bridge Report: 222,000 U.S. Bridges Need Major Repairs. https://www.artba.org/news/artba-2023-bridge-report-222000-u-s-bridges-need-major-repairs/ (accessed 25/04/2024).

Arup (2024) Boston Manor Viaduct. https://www.arup.com/projects/boston-manor-viaduct (accessed 16/04/2024).

BBC (2012) Japan Sasago tunnel collapse killed nine. https://www.bbc.com/news/world-asia-20576492 (accessed 17/04/2024).

Bridgeweb (2022) Results released from preliminary study into collapsed Norwegian bridge. *Bridge Design & Engineering.* https://www.bridgeweb.com/Results-released-from-preliminary-study-into-collapsed-Norwegian-bridge/8985 (accessed 17/04/2024).

CBDG (2007) *TG9 – Assessment of Concrete Bridges.* Concrete Bridge Development Group, Sandhurst, UK.

CNN (2021) A vital Memphis bridge shut down since May due to a structural crack has fully reopened. https://edition.cnn.com/2021/08/02/us/memphis-hernando-desoto-bridge-reopen/index.html (accessed 09/05/2024).

Cole G and Fish RJ (2022) *Highway Bridge Management.* ICE Publishing, London, UK.

Collins J, Ashurst D, Webb J, Sparkes P and Ghose A (2017) *Hidden Defects in Bridges – Guidance for Detection and Management.* CIRIA C764, CIRIA, London, UK.

Concrete Society (2010) *Technical Report TR22 – Non-Structural Cracks in Concrete*, 4th edn. www.concretebookshop.com (accessed 08/11/2024).

Concrete Society (2015) *Technical Report TR44 – The Relevance of Cracking in Concrete to Corrosion of Reinforcement*, 2nd edn. www.concretebookshop.com (accessed 08/11/2024).

Desnerck P, Lees JM, Valerio P *et al.* (2018) Inspection of RC half-joint bridges in England: Analysis of current practice. *Proceedings of the Institution of Civil Engineers – Bridge Engineering* **171(4)**: 290–302.

Dolling CN and Hudson RM (2003) Weathering steel bridges. *Proceedings of the Institution of Civil Engineers – Bridge Engineering* **156(1)**: 39–44.

Edwards T, Schofield M, Burdekin M and Neale B (2009) Failure of Roller bearings on the Thelwall Viaduct. *Proceedings of the Institution of Civil Engineers – Forensic Engineering. From Failure to Understanding* (January 2009), pp. 423–432.

FHWA (2022) National Bridge Inspection Standards – Federal Highway Administration. https://www.fhwa.dot.gov/bridge/nbis.cfm (accessed 09/05/2024).

FHWA (2023) National Bridge Inventory - Federal Highway Administration. https://www.fhwa.dot.gov/bridge/nbi/ascii2023.cfm (accessed 24/04/2024).

Freyssinet E (1923) *Le pont Candelier, pont-rail en béton armé sur la Sambre.* Annales des Ponts et Chaussées, Paris, France.

Gilbert M *et al.* (2022) *C800 Guidance on the Assessment of Masonry Arch Bridges.* CIRIA, London, UK.

Hambly E (1976) *Bridge Deck Behaviour.* Taylor and Francis, Abingdon, UK.

Harvey H and Harvey W (2016) *Hidden Defects 2016.* Bill Harvey Associates Limited. https://www.billharveyassociates.com/hidden-defects-2016 (accessed 18/04/2024).

HMSO (1945) *Memorandum (No. 577) on Bridge Design and Construction, Ministry of War Transport.* HMSO, London, UK.

LANTRA (2016) *National Sector Scheme 31 for the Bridge Inspector Certification Scheme.* https://www.lantra.co.uk/national-highway-sector-schemes-nhss/bridge-inspectors (accessed 26/04/2024).

Manning J (2024) Kendal road bridges could be knocked down following road closures. *BBC News.* https://www.bbc.co.uk/news/articles/cglk81dn3n5o (accessed 21/07/2024).

Merrison AW (1973) *Report of the Committee of Inquiry into the Basis of Design and Method of Erection of Steel Box-Girder Bridges.* HMSO, London, UK. https://www.istructe.org/getattachment/a1301a4d-7acb-4f3e-9623-4f8d6e1c91c1/attachment.aspx (accessed 28/01/2024).

National Archives (1956) National Interstate and Defense Highways Act (1956). https://www.archives.gov/milestone-documents/national-interstate-and-defense-highways-act (accessed 10/04/2024).

National Highways (2020a) CS 466 Risk Management and Structural Assessment of Concrete Half-Joint Deck Structures. *Design Manual for Roads and Bridges.* National Highways, Birmingham, UK.

National Highways (2020b) CS 468 Assessment of Freyssinet Concrete Hinges in Highway Structures. *Design Manual for Roads and Bridges.* National Highways, Birmingham, UK.

National Highways (2020c) CS 465 Management of Post-Tensioned Concrete Bridges. *Design Manual for Roads and Bridges.* National Highways, Birmingham, UK.

National Highways (2020d) CS 450 Inspection of Highway Structures. *Design Manual for Roads and Bridges.* National Highways, Birmingham, UK.

National Highways (2020e) CS 451 Structural Review and Assessment of Highway Structures. *Design Manual for Roads and Bridges.* National Highways, Birmingham, UK.

National Highways (2020f) CS 454 Assessment of Highway Bridges and Structures. *Design Manual for Roads and Bridges.* National Highways, Birmingham, UK.

Nicholls T, Armstrong C, El-Belbol S *et al.* (2015) Repairs to electro-slag weld defects on M4 Boston Manor viaduct. *Proceedings of the Institution of Civil Engineers – Bridge Engineering* **168(3)**: 259–272.

NTSB (2024) Collapse of the Fern Hollow Bridge, Pittsburgh, Pennsylvania, Jan. 28, 2022. https://www.ntsb.gov/investigations/AccidentReports/Reports/HIR2402.pdf (accessed 16/04/2024).

Pearson-Kirk D (2022) Post-tensioned bridges: Special inspections. In *Highway Bridge Management* (Cole G and Fish RJ (eds)). ICE Publishing, London, UK, pp. 113–132.

RAIB (2013) *Rail Accident Report: Partial Failure of a Structure Inside Balcombe Tunnel, West Sussex.* https://assets.publishing.service.gov.uk/media/5a81a888e5274a2e87dbecfb/130815_R132013_Balcombe_Tunnel.pdf (accessed 09/05/2024).

Rawlinson J and Stott P (1963) The Hammersmith Flyover. *Proceedings of the Institution of Civil Engineers*, London, UK.

Ricketts N (2017) *Railway Bridge Maintenance.* ICE Publishing, London, UK.

Roads.org (2020) Churchill Away. https://www.roads.org.uk/blog/churchill-away (accessed 06/05/2024).

Schacht G and Marx S (2015) Concrete hinges in bridge engineering. *Proceedings of the Institution of Civil Engineers – Engineering History and Heritage* **168**: 64–74.

Shave J and Bennetts J (2022) Assessment of load capacity of existing highway bridges. In *Highway Bridge Management* (Cole G and Fish RJ (eds)). ICE Publishing, London, UK, pp. 73–91.

Sholli S (2019) Liverpool viaducts brought down by catalogue of faults. *New Civil Engineer*, emap publications, London, UK, pp. 16–18.

Slade G (2023) Assessment of steel bridges – cracking the code. *The Structural Engineer* **101**(09): 14–20.

Smith A (1985) *A History of the County Surveyors' Society 1885–1985.* County Surveyors' Society, Shrewsbury, UK.

Stacy M, Denton S and Pottle S (2019) *Management of Safety-Critical Fixings. Guidance for the Management and Design of Safety-Critical Fixings.* CIRIA C778F, CIRIA, London, UK.

SteelConstruction.info (2018) Weathering steel. https://steelconstruction.info/Weathering_steel (accessed 25/04/2024).

Stroscio R, Malik A, Turlier F and Maynard M (2024) M5 Exe and Exminster viaducts – strengthening and safeguarding. *IABSE Symposium*, Manchester, UK.

Trout E (2010) The Development of Standard Prestressed Concrete Bridge Beams PDF. https://www.concretebookshop.com/the-development-of-standard-prestressed-concrete-bridge-beams-pdf-1517-p.asp (accessed 03/05/2024).

TSO (2007) *Inspection Manual for Highway Structures. Volume 1: Reference Manual and Volume 2: Inspector's Handbook.* TSO, London, UK.

UKBB (2005) *Management of Highway Structures, A Code of Practice.* TSO, London, UK.

UKMA (2024) Statistics. UK Motorway Archive. https://ukmotorwayarchive.ciht.org.uk/statistics/ (accessed 16/04/2024).

UKRLG (2016) UK Roads Leadership Group. https://ukrlg.ciht.org.uk/ (accessed 26/04/2024).

Wallbank EJ (1989) *The Performance of Concrete in Bridges*. HMSO, London, UK.

Watts M (2021) History of Hernando de Soto I-40 bridge in Downtown Memphis. https://eu.commercialappeal.com/story/news/local/2021/06/03/history-hernando-de-soto-40-bridge-downtown-memphis/5172758001/ (accessed 08/05/2024).

White House (2022) Remarks by President Biden after visiting the site of the collapsed Fern Hollow Bridge. https://www.whitehouse.gov/briefing-room/speeches-remarks/2022/01/28/remarks-by-president-biden-after-visiting-the-site-of-the-collapsed-fern-hollow-bridge/ (accessed 16/04/2024).

Richard Fish
ISBN 978-1-83608-559-1
https://doi.org/10.1108/978-1-83608-556-020251008
Emerald Publishing Limited: All rights reserved

Chapter 8
Understanding and managing risk

Most of the case studies in this book, whether on a collapse or a close call, were associated with a mismanagement of risk, either through a complacent unawareness, as at de la Concorde in 2006, or by apparently turning a blind eye, as at Polcevera in 2018. For every bridge stock, however, there will be other risks, such as oversights in the original design, detailing or construction which, more likely than not, will only manifest themselves once a bridge has been in service for several years. Bridge managers, however, have to deal with the stock which they have inherited and must manage risk accordingly (Mahmoud, 2019).

8.1.   Basic theory

The basic principles of risk analysis and management were set out in Chapter 1: those of addressing the likelihood of an event happening against the consequences of the occurrence. This is something that all of us must do several times a day, from taking decisions as to when it is safe to cross a road to walking downstairs with a cup of coffee and a pile of documents. These are known as *dynamic* risk assessments; those that rely as much on instinct or intuition rather than any degree of forward planning. Although these are qualities which also have a place in bridge management, this chapter will focus more on the need for preplanned risk analyses.

As with almost every subject in life, there are many levels of complexity which mean that things will always turn out to be more complicated than one first thinks. Risk management is no exception and there is certainly no shortage of reading material, training courses and the ubiquitous bespoke software packages to investigate, should the need arise to drill deeper into a specific method of analysis.

Although the topic of risk management in all sectors has grown significantly since the turn of the century, the late 1990s saw a growing emphasis on the need to better manage all infrastructure assets in order to prolong their working lives for as long as possible. Risk management is an implicit part of good asset management. The term risk management is believed to have first been coined in a business sense in 1993 when GE Capital appointed a 'Chief Risk Officer'. The first risk management standard was published jointly by the standards agencies in Australia and New Zealand in 1995. This was AS/NZS 4360 which was updated in 1999 and again in 2004 (Standards New Zealand, 2004). The twenty-first century awareness of risk and consequent expansion of risk management methodologies, however, are thought to have been triggered in part by the 9/11 terrorist attacks on the World Trade Center in New York, USA, in 2001 (BSI, 2020). This led to governments and businesses alike recognising the need to consider and evaluate risks which hitherto would have been unthinkable. In the UK, risk management was initially formalised through a guidance document, BS 31100, which was published in 2008, and was followed a year later by a standard, ISO 31000 (BSI, 2024).

With respect to bridges, risk management can be applied either to a single structure or to a bridge stock. For the former, this can help to determine the necessary holistic approach to dealing with a programme of planned maintenance. For the latter, although the risk analysis process will probably be far more granular, it should help to highlight the more problematic bridges in the stock and to prioritise interventions.

Before embarking on a risk register and getting absorbed in too much detail, a simple list of issues is a good place to start. The basic principles are then to objectively consider all aspects of that issue with respect to risk, an exercise which may even conclude in some cases that the risk is actually so trivial that it is not worth worrying about. For others, however, both likelihood and consequence need to be considered and 'scored'. A range of one to five for each is usually enough to give a sufficiently accurate picture.

Setting these out in a matrix, and inserting the product of the two numbers, gives scores as shown in Figure 8.1. In this example, the highest risks have been chosen as those with a score greater than 20, with medium-level risks sitting between 10 and 16. A score of anything less than ten may be considered as a low-level risk, although the option used occasionally is to consider scores of one or two as 'negligible' risks (Shetty *et al.*, 1996). Those demarcations can be set at any figure and can also be colour coded in a traffic light, red-amber-green (RAG) arrangement to help explain the risk approach to non-engineering managers or politicians.

It is suggested that there should be some reflection on the outcomes from a risk analysis such as this, both for individual risks and where the RAG thresholds have been set, rather than jump to immediate conclusions.

Having produced a prioritised register of risks, there needs to be further work as far as risk management is concerned, to make it much more than just a list. As well as identifying which risks should be given the highest priority, each should be allocated possible mitigation measures both in the immediate term and as a longer-term strategy for closing them out.

Another important factor is the need for ownership of risks. In a larger organisation, individual members of a team all need to fully understand those risks for which they are responsible and to know when, and to whom, as necessary, the risk should be escalated. The risk register also needs

Figure 8.1 Example of a simple risk matrix (author's own)

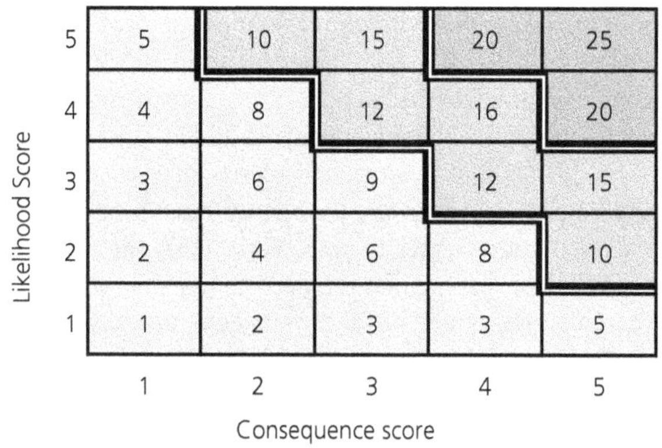

Likelihood Score					
5	5	10	15	20	25
4	4	8	12	16	20
3	3	6	9	12	15
2	2	4	6	8	10
1	1	2	3	3	5
	1	2	3	4	5

Consequence score

to be a live document. It should be central to all bridge management decisions and reviewed on at least a monthly basis, using data from monitoring regimes and/or inspections.

In practical terms, and recognising that no risk can ever be completely eliminated, the accepted approach is to consider risks being 'as low as reasonably practicable' (ALARP). The term is thought to have been first used by the UK's Health and Safety Executive (HSE) in 1992 when guidance was issued for addressing risk in gas storage containers (HSE, 1992). The principle is illustrated in Figure 8.2 which shows how unacceptable risks can be driven down, not to an area where they might be deemed acceptable, or even non-existent, but to a position of an intermediate situation: ALARP.

The ALARP principles can be applied throughout the various aspects of risk with regard to bridge management which are covered below.

Figure 8.2 The ALARP principle (Author's own)

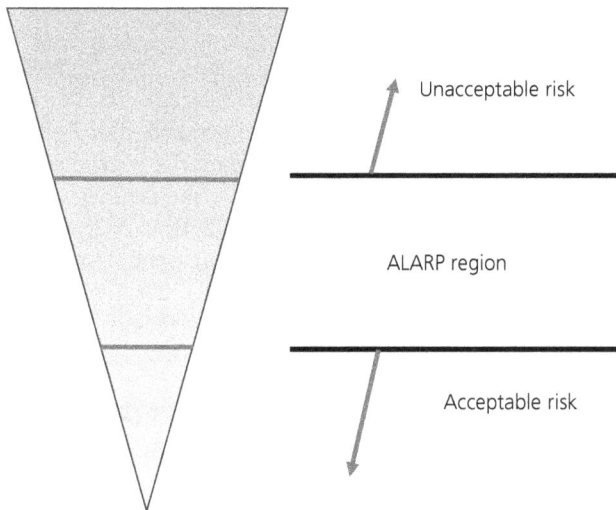

8.2.   Management of substandard structures – background

The most obvious application of risk management is in the management of substandard bridges. These are structures that are known to have at least one shortcoming; usually when a structural assessment has revealed a lower load-carrying capacity than national standards require. The now codified concept of managing such structures first came about in the UK somewhat indirectly as a consequence of a European Union (EU) directive from 1984 which sought to harmonise gross vehicle and axle weights for heavy goods vehicles across all countries in the EU, coming into force in 1999 (OJEC, 1984). In the UK, this led to a change in the Construction and Use Regulations such that, in simple terms, the maximum gross vehicle weight increased from 38 to 40t and the maximum axle weight from 10.5 to 11.5t (C&U Regs, 1986). Although these increases may seem relatively modest, for small- to medium-span bridges, an almost 10% increase in axle live load was not insignificant in terms of both bending and shear effects.

In order to meet the EU Directive, the UK government instigated and funded a programme of bridge assessment and strengthening (BASP), beginning in 1987 (Das, 1996). For local highway authorities, funds had to be bid for, against submitted programmes of assessment and strengthening, in their annual Transportation Policies and Programme (TPP) submissions (later replaced by an initially annual and latterly five-yearly Local Transport Plan (LTP)). As part of the BASP, several new standards were introduced, notably BD 21/93 (later revised in 1997 and again in 2001): The Assessment of Highway Bridges and Structures (National Highways, 1997).

As the programme gathered momentum, it became clear that there was bound to be a considerable lag between a bridge having failed its assessment and having a strengthening (or even replacement) project fully implemented. Irrespective of the availability of funding, this was not an easy step as there needed to be feasibility studies and options appraisals well before any detailed design work could begin. Similarly, while the programme was, in theory, 100 per cent government funded, there were limitations on how quickly grants could be allocated. The only solution in addressing a failed assessment during this period was for highway authorities to impose formal interim measures, such as temporary weight or width restrictions. On major roads, however, such an action would have had serious socio-economic impacts on local communities as well as unrealistic detours on more strategic routes.

It was the (then) County Surveyors' Society Bridges Group and the Highways Agency which set up a working group to consider how theoretically 'failed' bridges should be managed. Their proposals were accepted and originally published as an advice note, BA 79/98 The Management of SubStandard Highway Structures, but subsequently updated as a standard with the same title in 2006 as BD 79/06 and later revised as BD 79/13 in 2013 (National Highways, 1998). A complete revision of the *Design Manual for Roads and Bridges* (DMRB) in 2020 saw BD 79 being replaced by the current standard, CS 470 (National Highways, 2020a). Central to all these standards was the management of risk by setting out a process that not only gave specific guidance on how to manage the issues within a short-term horizon but also left sufficient room to be able to take a pragmatic approach in arriving at a workable, long-term solution.

The impact of having the original advice note in 1998 cannot be overstated. For the first time, there was a framework within which to exercise the engineering judgement that bridge managers had for many years been using to inform their decisions: both consciously and subconsciously. The advice introduced concepts such as 'Immediate Risk Structures' and the suite of 'Interim Measures' that could legitimately be put in place (Cole, 2000). As will be seen below, such terms remain in use within the latest CS 470 standard.

As well as relying on the initial DMRB advice, many highway authorities developed similar approaches but with greater emphases on local issues – for example, the organisation covering all of the boroughs in London, UK (London Bridge Engineering Group (LoBEG)) produced guidance which added other factors to their prioritisation methodology, including the social impact of having a restricted bridge on the network (Shetty *et al.*, 2000). Similarly, a model was also developed for the UK's motorway and trunk road network, taking into account subtly different criteria (Ives and Jandu, 2005), as well as a more pragmatic approach taken in Scotland (Brodie, 1996).

Although the UK government's original expectation may have been that the bridge assessment and strengthening programme should have been completed by the time the EU directive was implemented, this has proved not to have been the case. In fact, bridge assessments and the management of substandard structures have become a major part of the bridge manager's ongoing day-to-day workload since the turn of the century. Section 8.6 below examines the need to consider risk in bridge assessments in more detail.

8.3.   Management of substandard structures – current UK practice

Although CS 470 follows the same principles as had been established by the original BA 79 advice note and BD 79 standard, it is much more prescriptive, especially in terms of record keeping and the approval processes. Clauses 2.1 and 2.2 from CS 470 are given in Table 8.1 and summarise the key desired outcomes very well.

Table 8.1 Clauses from CS 470 Management of Sub-Standard Highway Structures (courtesy of National Highways)

2.1	Provisionally substandard structures and substandard structures shall be managed by assessing the risks to public safety associated with their continued use and imposing appropriate interim measures when necessary.
2.2	Load mitigation interim measures shall be urgently imposed on immediate risk structures.

Provisionally substandard bridges might include those that have been identified by way of an inspection, including obvious examples such as damage due to a scour event or from a vehicle impact, as well as any unexpected degree of deterioration. They can also be those in which an initial structural assessment gives results that are unlikely to be improved upon through a more rigorous analysis. As the name implies, an immediate risk bridge is one that presents an unacceptably high safety risk to the public, either on or below the structure. Against a number of given criteria, CS 470 states that an immediate risk structure can be identified by any *competent* individual.

The management methodology in CS 470 for dealing with a provisionally substandard structure is set out in Figure 8.3[1] in which risk can be seen to be the key term driving the decision-making process.

If a provisionally substandard bridge is deemed to be low risk, then the next consideration in the process is to determine whether it is 'monitoring appropriate'. To satisfy this requirement, several criteria must be met, including the degree of any pre-existing structural distress and also that any further deterioration is going to be progressive rather than leading to a sudden catastrophic failure. Another requirement is that the monitoring option must be both 'meaningful and effective'. It is also important that any monitoring must be objective and not simply visual, probably utilising some form of basic instrumentation such as strain gauges or crack width tell-tales fixed to the bridge, or more sophisticated techniques such as acoustic emission monitoring (Cousins and Anderson, 2022). A vital part of a monitoring plan is to establish trigger levels from the outset. Examples might include introducing a limiting dimension of a crack width or length, or when a deflection reaches a predetermined value. The monitoring plan should also list actions that will need to be taken once a trigger level has been reached.

The concluding boxes of Figure 8.3 refer to the option of 'interim measures'. While monitoring itself is an interim measure, others are mainly related to some form of load mitigation. The obvious choice here is the imposition of a weight restriction or by introducing lane restrictions, either to reduce the total volume of traffic loading or to restrict vehicles from particularly vulnerable elements such as cantilevers. Weight and lane restrictions are also governed by a DMRB standard

---

[1] This is Figure 4.1N2 from CS 470. See also CS 470 Figure 5.1N2 for the management process of substandard structures.

Figure 8.3 Management process for provisionally substandard structures (courtesy of National Highways)

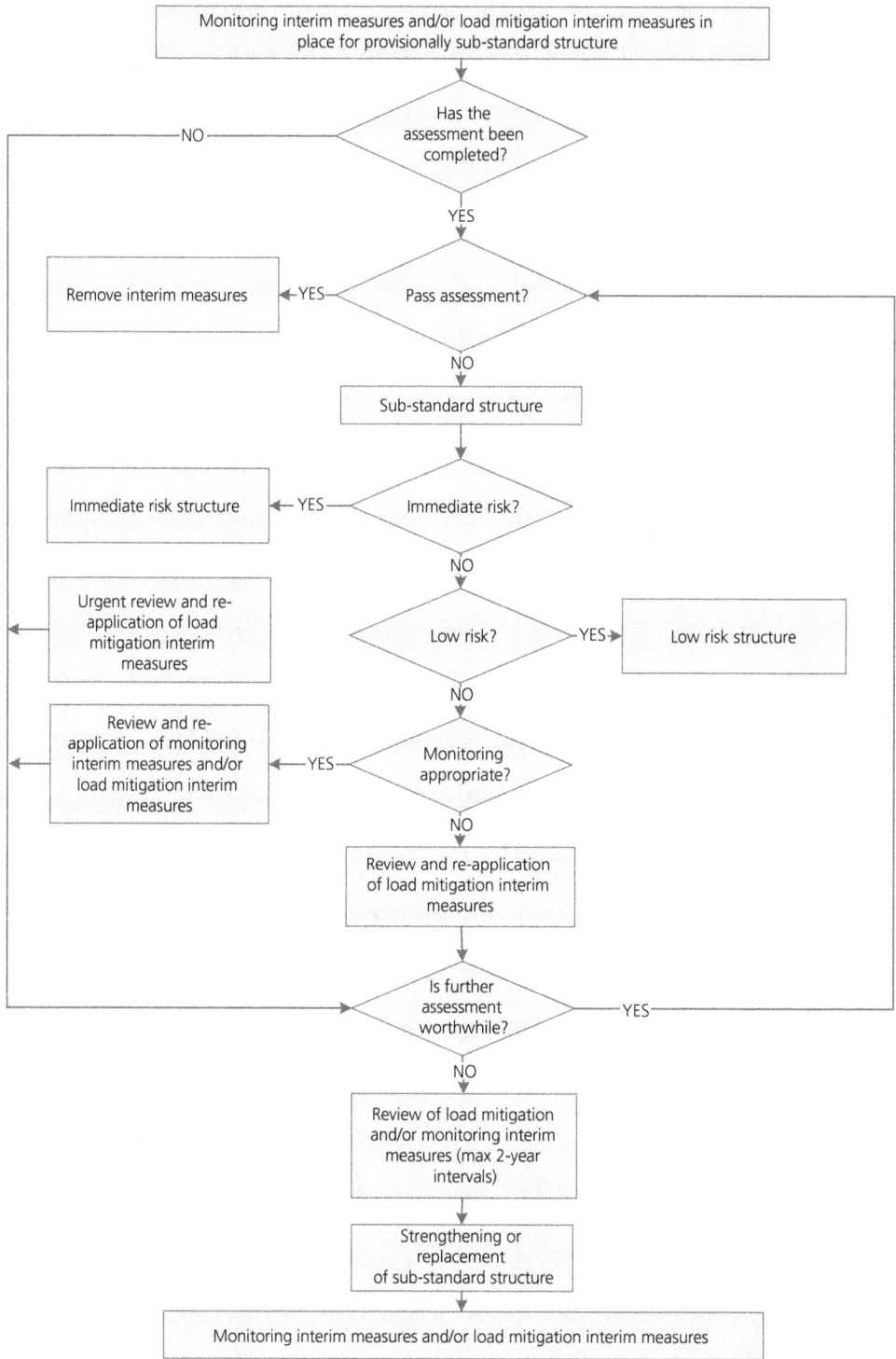

(National Highways, 2020b). Other acceptable interim measures would be to install some form of temporary propping or to use a temporary bridge either over or alongside the bridge.

Irrespective of the interim measures chosen, they themselves must be monitored and maintained in order to ensure their continued efficacy. CS 470 also requires that an emergency response and communications plan is in place ready to be activated at short notice as and when needed. Interim measures must also be subject to a regular review process at predetermined intervals.

As earlier chapters have shown, the loss of records during a bridge's life can be a contributory factor to a failure. In the case of an already substandard structure, which could have been under interim measures for several years, it is essential that record keeping is given a high priority. For this reason, CS 470 is prescriptive on this point in order to ensure that all decisions are recorded, and approvals obtained, from beginning to end of the substandard structure management process; from the initial identification of a provisionally substandard bridge to the end point of when interim measures can be removed. To this end, the standard contains several appendices which set out the approval requirements through the various stages of the process.

Before closing this section, it should be recalled that Chapter 1 introduced the subject of precursor events. These may arise at any stage of a bridge's life, whether during design, construction, its working life or even demolition. The link between the identification of precursors and the current UK practice with regard to the management of substandard structures, as exemplified by CS 470, is far from tenuous. The key player in both is the bridge manager who must have the vigilance to recognise precursors when they arise and the confidence to work to the standard to defuse them. There is also little room for the characteristic of being risk averse, as this can result in a syndrome of paralysis by overanalysis. The far better attribute is to be risk aware; to take action as and when the need arises.

8.4.   Risk in bridge design

In considering the possible root causes of bridge collapses, Chapter 3 highlighted design errors, both in concept and in the misapplication of codes and standards, as one of eight generic reasons why a bridge might fail.

In order to ensure that the risk of errors in both conceptual and detailed design is minimised, the UK has a rigorous technical approval process in place which, if followed, should help to eliminate such risks. As noted in the case study in Chapter 5 on the 1970 Cleddau steel box-girder bridge collapse, the concept of technical approval, and the need for certified design and checking, came about as a result of recommendations from the committee of inquiry into the collapse (Merrison, 1973).

The process has been greatly enhanced over the intervening years, culminating in the current standard, CG 300 Technical Approval of Highway Structures (National Highways, 2021). Explicit in this procedure is the need for a Technical Approval Authority (TAA), usually the client organisation, the bridge owner. The TAA must be confident that they are going to have the bridge they want in terms of concept and that its design will be fully compliant with current standards. This is achieved by the designer completing an 'approval in principle' (AIP) form which must firstly state why the chosen design has been proposed, evidenced by a costed options appraisal, often as an appended report. The TAA must be sufficiently competent to challenge assumptions made in the AIP, a point made elsewhere on the need for the 'intelligent client' concept.

There are four possible categories of structure in CG 300, 0 to 3, depending on the complexity. These categories also dictate the level of design check required. The simplest, *categories 0* and *1*, can be checked within the same team as the designer. *Category 2* designs can be checked within the same organisation but by a different team. *Category 3* designs must be the subject of a completely independent check by a different organisation. Irrespective of the category, design and check certification is also a requirement. The checking process is equally as important as the design and should be given a similar level of emphasis and suitably resourced. The design check is the most obvious way of reducing risks arising from bridge design.

Although it may at first sight appear that this process will constrain designers and restrict opportunities for innovation, this is not the case. For complex structures, the facility exists to depart from a DMRB standard and even in special cases to consider aspects that are not covered by a standard. Such departures need to be incorporated in the AIP and approved by the TAA.

In the UK, CG 300 requires that the AIP is signed by a professionally qualified engineer, chartered through either the Institution of Civil Engineers or the Institution of Structural Engineers. Such qualifications apply not only to those submitting the form but also the TAA. Should the highway authority client have no-one with such qualifications then the TAA role may be subcontracted to another party, normally a consulting engineer not connected with the project. This, however, is a far from ideal situation; far better to have qualified in-house expertise who can fulfil the role of an 'intelligent client'.

8.5.   Risk in bridge construction

Accepting that there are innumerable risks on any construction site, the majority of those relate to health and safety considerations which, as important as they are, are outside the scope of this book. That said, in the UK, the Construction (Design and Management) (CDM) Regulations (HSE, 2015) place considerable responsibility on the role of the 'Principal Designer'. Those responsibilities include ensuring that coordination and communication between all parties is given a sufficiently high priority.

Aside from CDM, the risks to be considered here are those which could lead to a structural failure and catastrophic collapse, such as that of the 1970 West Gate Bridge collapse, described as a case study in Chapter 5. Here, as with many others, it was mostly the human factors that proved to be the root cause. These included not only a failure to understand structural effects in the erection condition but also, perhaps more worryingly, basic communication issues together with an apparently strong dose of complacency.

Another case study already cited is that of the Chirajara Bridge collapse in Colombia in 2018, the investigation of which is referenced in Chapter 4 and, although fundamentally caused by an error in design, led to the collapse during construction.

As was the case at West Gate, it is the temporary condition of a bridge under construction, or the temporary works required to build it, that present the highest risks. Irrespective of the disclaimer in the opening paragraph of this section, it is probable that temporary works failures lead to far more serious incidents on site than any other root cause.

Temporary works will also include lifting or erection processes either using cranes or bespoke launching gantries, as well as the permanent works in a temporary condition. As Table 4.1 shows, it is India that has the worst fatal collapse record of any country in this century and many of these have been during construction, most recently when a railway bridge under construction at Sairang failed, killing 26 workers (BBC, 2023) and when a launching gantry collapsed on an under-construction flyover on a new expressway in Maharashtra, killing 20 (Sarkar, 2023), both in August 2023. Examples of such failures are not uncommon closer to home: in February 2024, two workers were killed when a tandem crane lift of a new steel arch bridge failed in Lochem in the Netherlands (Johnson, 2024).

Although the above are just examples, they serve to show that risks in temporary works will generally have significant consequences. Although there are many others which could have been featured in passing, or as more detailed case studies, these should serve to demonstrate that temporary works design should be given an equivalent emphasis to that required for permanent works. For this reason, in the UK, CG 300 includes AIP template forms that must be submitted for some specific temporary works designs. Also in the UK, temporary works now has its own professional body, the Temporary Works Forum, covering the interests of both designers and those of specialist equipment suppliers (TWF, 2024).

Before moving on to other aspects of bridge management in the round, a salutary reminder of actual risks in bridge design and construction is to be found in the October 2018 CROSS newsletter (CROSS-UK, 2018). Bearing in mind that this was the year of the Polcevera collapse, the following comments (based on an anonymised report) on a UK bridge construction site demonstrate a level of concern which it must have been hoped should never be the case.

> It is worrying that we should come across such a large and complex structure being, apparently, so poorly designed and executed. Especially so at a time when we are all so focused on quality of design and construction and after there have been recent major bridge failures.

> Key factors in avoiding such issues are to use experienced Chartered Engineers for both design and construction, and to have regular and effective communication between all parties. (CROSS-UK, 2018, p. 2)

Sadly, does this not suggest that the lessons of the past have not been sufficiently well communicated to the present?

8.6. Risk in bridge assessment

Arguably of more importance than risks in design are the risks associated with the process of structural assessment. Bridge designers have the advantage of starting with a clean piece of paper, a blank canvas, on which to develop their ideas. The assessment engineer, however, is presented with a structure often about which very little is known and, hence, they are faced with a number of challenges around the assumptions they must make, and the level of detail that they might need to determine, before having the confidence to reach a decision on structural capacity (Shave and Bennetts, 2022).

Referring to the UK bridge assessment and strengthening programme in the 1990s, there was often a clear aversion to risk when it came to an assessment outcome, mostly driven by fee competition. Consulting engineers asked to bid for a tranche of assessments would often be asked to submit a lump sum price which, in order to win the commission, they knew had to be as low (and, for them, as risk free) as possible. This was not only a race to the bottom but also counterintuitive: was it not worth a bridge client spending more on an assessment in order to squeeze as much residual life from a bridge such that unnecessary strengthening or replacement could be prevented? Sadly, procurement specialists, and corporate diktats, erroneously focusing on perceived value for money principles, seemed to have a greater clout than those espoused by engineering clients.

In such cases, a phenomenon of mutual risk aversion was occasionally to be found, neither client nor consultant having the courage to delve deeper into an assessment 'failure' and, therefore, missing the opportunity to extend a bridge's service life with minimal risk. Ironically, it is only when funding for strengthening or replacement became harder to find (*c.* 2008 onwards), did such exercises become more commonplace.

Possibly as a result of a risk-averse approach having been taken, and with its inherent conservatism, it seems that there is as yet no evidence of a bridge failure that has arisen solely as a result of an incorrect assessment result.

8.7. Risk in bridge inspection

As I have said many times in this book, arguably the most important tasks in managing an existing bridge stock are inspections. These form the bases of all data and information gathering upon which decisions on whether (and when) maintenance interventions are going to be needed are based. Risks here are numerous. An inspector may miss a significant defect, fail to objectively measure extents, or misinterpret cause and effect – for example, has a crack with vegetation growing within it been caused by the expansive nature of the roots or has it been formed by another structural mechanism to provide that vegetation with a space to grow?

Very few, if any, bridge management procedures have the luxury of having all (or even a sample) of inspections checked by an independent party. Risk, therefore, can only be addressed by assessing the competence of inspectors. As noted in Chapter 7, proving that level of competence by a certification scheme is vital in removing such risks (LANTRA, 2016).

8.8. Risk in bridge management

Although most of the above sections have encompassed bridge management in the round, they are all tools at the disposal of a bridge manager to help them in their job. It is the bridge manager's responsibility, however, to combine all of the issues they face, within the constraints placed upon them, to deliver a bridge stock that is as safe and risk free to the travelling public as it possibly can be.

To do this is no small achievement. Irrespective of the support, or otherwise, they receive from their senior management, bridge managers are constantly juggling conflicting pressures to keep a network unrestricted and to provide a reassurance that the risks they are dealing with are negligible. As also noted in Chapter 1, but well worth repeating here, the late eminent bridge engineer Dr Bill Harvey has remarked that there should be 'a constant low level of anxiety' for all who are charged with managing bridges (Harvey, 2021).

An implicit risk in bridge management, however, lies in the capacity, capability and, above all, competence of a bridge manager and his team. These should be essential characteristics, but can they be quantified to prove that there are the right number of professionally qualified individuals, with an appropriate level of training and experience, to be able to inform a bridge manager who must make those difficult judgement calls?

Risk, therefore, is a constant. For a bridge manager, perhaps the primary quality is one of being risk aware; to have an 'eyes wide open' approach, to be vigilant in all aspects of bridge management and to go nowhere near complacency, in order to ensure that a bridge stock remains as risk free as possible.

8.9. Risk case study 1: Francis Scott Key Bridge, Baltimore, USA

With regard to conceptual design, it is worth reflecting on the immediate media and public reaction to the March 2024 collapse of the Francis Scott Key Bridge in Baltimore, USA (The Guardian, 2024). That such a major bridge could catastrophically fail was truly shocking to the uninitiated observer, with reactions much along the lines of, 'Why should the whole bridge collapse when just one support was taken out?' or 'Could it not have been designed for such an eventuality?'

Opened in 1977, the 2.6 km bridge had main spans of a continuous steel arch through truss with a central clear span of 366 m, making it the world's third-longest span of this form (Figure 8.4). At about 1.30 am on 26th March 2024, one of the central span's supports was hit by a 300-m-long errant container ship, the MV *Dali*, as it was leaving the port (Figure 8.5). The collision took out the support, leading to the catastrophic collapse of the truss spans and claiming the lives of six workers who were carrying out repairs to the carriageway. The support structure is shown in Figure 8.6 and serves to illustrate the relative lightness of the design (in keeping with the superstructure). As with the similar collapse of the Sunshine Skyway Bridge, noted in Chapter 5 as a case study, the NTSB treated this event as a marine incident investigation but were able to issue a preliminary report just some six weeks after the collapse (NTSB, 2024).

Returning to public reaction: yes, a bridge could be designed to remain standing with one support removed but at such an additional cost, either for a ridiculously stiffer truss or a completely

Figure 8.4 Francis Scott Key Bridge (public domain, author unknown)

Figure 8.5 Francis Scott Key Bridge after the collapse (public domain, NTSB)

Figure 8.6 The surviving main support structure for the Francis Scott Key bridge. The opposite support was taken out by the MV Dali (courtesy of Wikimedia Commons. This file is licensed under the Creative Commons Attribution 2.0 Generic licence (Attribution 2.0 Generic license – Creative Commons))

different form (for example, by having a much greater span, such as a suspension, or cable-stayed, bridge). Convincing politicians in the early 1970s that the additional cost could be justifiable, on the basis of an incident such as occurred, would have been challenging to say the least.

More importantly, however, is the need to acknowledge that risk (in terms of both likelihood and consequence) changes over time. A 1970s risk assessment into a possible ship collision would be very different to one in the 2020s, if only for the reason that the size of vessels and the volume of merchant shipping have increased by factors of about seven over that period (transportgeography. org, 2024). That suggests that the risk has grown by at least the same figure: in plain sight over the last 50 years. Noting that the Sunshine Skyway collapse in 1980 was the trigger for all new designs to have ship protection, it is likely that the 2024 Francis Scott Key collapse will be the trigger for a long-overdue review, not only for new designs but also for existing bridges where the risk of ship impact is high (Caprani, 2024).

A further point to note, as referenced in the I-35W case study in Chapter 6, is that the Francis Scott Key Bridge was another example of the American love affair with the steel truss. As a result, it was also classified as a 'nonredundant steel tension member' bridge. Or more scarily, the previous term: fracture critical. Table 7.1 shows that 42391 US bridges are structurally deficient; the fact that within that figure there are 17468 fracture critical bridges (Weiss, 2024) just emphasises the scale of the problem facing the USA; one which must have a very large number of risks of the highest priority.

8.10. Risk case study 2: Florida International University footbridge, Miami, USA

On 15th March 2018, a recently erected span of a somewhat unconventional pseudo-truss concrete footbridge over SW 8th Street in Miami, Florida, USA, collapsed onto live traffic lanes below (NTSB, 2019). Five people died in vehicles under the bridge and a construction worker on the bridge was also killed (Figure 8.7). This failure was also covered by SCOSS (as it was known at that time) in the UK (CROSS-UK, 2020).

The key lesson to be learnt from this collapse was that precursor events, and their consequent risk, were not only obvious but also ignored. Almost as soon as the first span of the bridge had been positioned on its substructure, and hence carrying its own dead load, cracks were noticed at some of the truss nodes. Over the next fortnight the cracks continued to propagate, and various explanations and corrective actions were considered. The Engineer of Record,[2] however, is noted

Figure 8.7 FIU bridge collapse (courtesy of CROSS-UK)

[2]An Engineer of Record (a US term) is the person responsible for the design phase of a project and for seeing that a structure is built according to the design.

in the NTSB report as stating that he '...*repeatedly indicated that the cracks were of no safety concern...*'. (NTSB, 2019, p. XV of the Executive Summary). On the day of the collapse, 19 days after being erected, and with crack widths having grown to up to 20 mm, one of the prestressed diagonals was being retensioned in an attempt to stabilise the structure. Although this was later proven to have been a mistaken intervention which actually triggered the failure, it seems probable that the distress in the bridge meant that a catastrophic collapse was inevitable.

Risks here were clear and obvious. It is hard to comprehend the thinking and decision making of those responsible for the bridge during those 19 days. An obvious mitigation would have been to close the road below as soon as the unexplained cracking had been identified. Instead, complacency and risk-blindness appears to have prevailed, with tragic consequences.

8.11.   Risk case study 3: Bridge 88, Stewarton, Scotland, UK

This was a typical, modest single-span bridge of conventional form carrying the Glasgow to Carlisle railway over the A735 road just south of Stewarton in Scotland. It was a half-through design with main supporting elements of three wrought-iron plate girders. It had a skew of 39° and a square span of 9.2 m. The bridge dated from the 1870s and originally carried twin tracks. It was reduced to a single track in the 1970s.

Early in the morning of 27th January 2009, a freight train consisting of a locomotive and ten tank wagons (carrying gas oil, diesel and kerosene) was crossing the bridge when it collapsed, and the last six wagons derailed. Although there were no injuries, there was a significant environmental incident with much of the fuel leaking into watercourses.

A highly forensic investigation and detailed report (RAIB, 2010) found that the eastern and central girders had failed catastrophically. It also noted that the webs were so heavily corroded that the shear affects could no longer be resisted and this had triggered the collapse. The bridge's structural form and detailing had meant that there was a hidden 'trap', meaning that water had been allowed to pond against web plates. Corrosion had taken place over the years, such that the section loss had become very significant and even to the extent that holes had formed in these areas of the webs.

The bridge had been inspected in accordance with national standards at the requisite intervals and had also been assessed. The assessment, however, had assumed full section thicknesses with no reduction due to corrosion. This was also an oversight which could have had serious consequences as all assessments should be based on actual dimensions, including plate thicknesses. A Detailed Examination report (the equivalent of a Principal Inspection) in 2003 had highlighted a potential problem with section loss but this was not acted upon. Another factor related to maintenance work which had been carried out in 1987 when track ballast was removed, exposing evidence of historical web corrosion. No repairs nor repainting, however, had been undertaken at that time. Bridge 88 had also been identified for replacement. In fact, the new bridge deck had already been fabricated and installation had been planned just six days after the collapse. The replacement plans had not only been the focus of attention, with any interim intervention probably considered unnecessary, but also, as a result, perhaps introduced a trace of complacency.

The term 'sweating the asset' comes to mind in this case. The owners must have been sufficiently aware of the risks, in that they were about to replace the bridge, but were they risk blind in terms of the rate of deterioration? Throughout the whole of the bridge management process there had been opportunities to raise awareness of the potential risks, but the only action taken seems to have been to focus on replacement.

8.12. Risk case study 4: Menai Suspension Bridge, Wales, UK

Opened in 1826 and with a main span of almost 177 m, the Menai Suspension Bridge connects mainland Wales with the Isle of Anglesey (Figure 8.8). It was designed by Thomas Telford, the first president of the Institution of Civil Engineers, and is a Grade I listed structure. The bridge has two traffic lanes and cantilevered footways outside the line of the suspension system. As one of only two bridges spanning the Menai Strait, it remains a vital strategic route.

Bearing in mind that two contemporary suspension bridges (Broughton and Great Yarmouth, which opened at about the same time) are featured as nineteenth-century collapse case studies in Chapter 2, it is worth reflecting on the sound management of Menai as it is approaching its bicentenary. There have, of course, had to be several significant maintenance interventions in that time, most notably in 1938 when a complete replacement of the suspension system and deck was undertaken (Day, 2012). Another major intervention came between 1988 and 1991 as part of the bridge assessment and strengthening programme noted in Section 8.2 above. At this point, earlier concerns about potential brittle failure of the hanger socket castings were reviewed and several hangers removed and tested; all failed at the casting rather than the hanger itself. A number of the most highly loaded hangers were then replaced, meaning that the bridge was once again certified as being adequate for full HA traffic loading (i.e. the standard UK design loading).

In January 2022, a detailed study of earlier works concluded that potential brittle failure remained a significant risk, and a 7.5 t weight restriction was imposed, pending a more detailed analysis of possible failure mechanisms. That analysis showed that, should a hanger fail, the majority of its

Figure 8.8 Menai Suspension Bridge (courtesy of Structurae)

load would jump to adjacent hangers which themselves would become overloaded and initiate an unzipping of the whole suspension system with catastrophic consequences (Wood *et al.*, 2023).

Despite concerns about the traffic disruption, the impact on the local economy and the reduced resilience of the connectivity between Anglesey and the mainland, in October 2022, the decision was taken to close the bridge completely. Emergency works then took place to install secondary, fail-safe hangers which allowed the bridge to be reopened on 1st February 2023. At the time of writing, permanent works are currently being prepared.

The difference in the approach to risk between this case study and that of the FIU is clear to see. At Menai, as soon as the risk had been noted and effectively quantified, some tough (and to many, unpopular) decisions were taken. A collapse was avoided, and an iconic bridge preserved.

8.13. Conclusion

The management of risk is always going to be a judgement call, one that requires both an awareness of risks and an objective and analytical approach to take appropriate action. It is essential that any risk register remains a live document and must include risk mitigation measures and absolute clarity as to who is responsible for managing each risk. As noted with the Francis Scott Key Bridge, risk evaluation can change with time, and this should be part of any review of more long-standing risks.

Most importantly, a bridge manager must remain risk aware but cannot afford to be either risk averse or risk blind. A professional approach to risk is one which touches on engineering ethics, which will be covered in the next chapter.

REFERENCES

BBC (2023). India bridge collapse: At least 26 killed at construction site. https://www.bbc.com/news/world-asia-india-66590539 (accessed 19/05/2024).

Brodie A (1996) The Assessment and Strengthening Programme. The Scottish Office Approach: "The Pragmatic Engineer". In *Bridge Modification 2: Stronger and Safer Bridges*. Thomas Telford, London, UK, pp. 9–14.

BSI (2020) Risk Management: Manage Your Risks. https://www.bsigroup.com/globalassets/localfiles/en-in/certification/iso-31000/iso-31000-risk-management.pdf (accessed 13/05/2024).

BSI (2024) ISO 31000 Risk Management. https://www.bsigroup.com/en-IN/ISO-31000-Risk-Management/ (accessed 13/05/2024).

C&U Regs (1986) The Road Vehicles (Construction and Use) Regulations 1986. https://www.legislation.gov.uk/uksi/1986/1078/contents/made (accessed 13/05/2024).

Caprani C (2024) The Baltimore bridge collapse: A vital lesson for structural engineers. *The Structural Engineer* **102(5)**: 40–41.

Cole G (2000) Managing sub-standard bridges. *4th International Conference on Bridge Management*, Thomas Telford, London, UK.

Cousins D and Anderson M (2022) Testing and monitoring. In *Highway Bridge Management* (Cole G and Fish RJ (eds)). ICE Publishing, London, UK, pp. 93–112.

CROSS-UK (2018) Quality of design and construction of a major bridge structure. *CROSS-UK Newsletter 52*. https://www.cross-safety.org/sites/default/files/2018-10/cross-uk-newsletter-52.pdf (accessed 22/05/2024).

CROSS-UK (2020) Lessons learned from the 2018 Florida bridge collapse during construction. https://www.cross-safety.org/sites/default/files/2020-12/lessons-learned-2018-florida-bridge-collapse.pdf (accessed 22/05/2024).

Das P (1996) Bridge Assessment and Strengthening – Policy and Principles. In *Bridge Modification 2: Stronger and Safer Bridges*. Thomas Telford, London, UK, pp. 1–8.

Day W (2012) Menai Suspension Bridge: A history of maintenance and repair. *Proceedings of the Institution of Civil Engineers – Engineering History and Heritage*. https://www.icevirtuallibrary.com/doi/10.1680/ehah.2012.165.1.9 (accessed 24/05/2024).

Harvey W (2021) Some thoughts on competence. *Sixty-sixth Meeting of the Bridge Owners Forum* (January 2021). https://www.bridgeforum.org/bof/meetings/bof66/BOF66%20-%20Harvey%20-%20Competence%20note-202101a.pdf (accessed 21/05/2024).

HSE (1992) Safety Report Assessment Guide: Methane Gas Holders. https://www.hse.gov.uk/comah/sragwmh/criteria/section1.htm (accessed 23/05/2024).

HSE (2015) Construction (Design and Management) Regulations 2015. https://www.hse.gov.uk/construction/cdm/2015/index.htm (accessed 06/06/2024).

Ives D and Jandu A (2005) Maintenance prioritisation of highway structures. *5th International Conference on Bridge Management*, Thomas Telford, London, UK, pp. 433–442.

Johnson T (2024) Under construction road bridge in Netherlands collapses killing two Bam sub-contractors. *New Civil Engineer*. https://www.newcivilengineer.com/latest/under-construction-road-bridge-in-netherlands-collapses-killing-two-bam-subcontractors-23-02-2024/ (accessed 19/05/2024).

LANTRA (2016) National Sector Scheme 31 for the Bridge Inspector Certification Scheme. https://www.lantra.co.uk/national-highway-sector-schemes-nhss/bridge-inspectors (accessed 26/04/2024).

Mahmoud K (2019) *Risk-Based Bridge Engineering*. CRC Press, London, UK.

Merrison AW (1973) *Report of the Committee of Inquiry into the Basis of Design and Method of Erection of Steel Box-Girder Bridges.* HMSO, London, UK. https://www.istructe.org/getattachment/a1301a4d-7acb-4f3e-9623-4f8d6e1c91c1/attachment.aspx (accessed 28/01/2024).

National Highways (1997) BD 21/97 AT01 The Assessment of Highway Bridges and Structures. https://www.standardsforhighways.co.uk/search/6ea33d54-3c11-4dd0-aa6f-ed17981d825d (accessed 13/05/2024).

National Highways (1998) BA 79/98 The Management of Sub-Standard Highway Structures. https://www.standardsforhighways.co.uk/search/b3a3a15f-cac1-4ded-b7c7-dfe01a75eaba (accessed 13/05/2024).

National Highways (2020a) CS 470 Management of Sub-Standard Highway Structures. https://www.standardsforhighways.co.uk/search/8d9db6a3-55e2-4947-855a-98ae3db77fc5 (accessed 13/05/2024).

National Highways (2020b) CS 454 revision 1. https://www.standardsforhighways.co.uk/search/html/96569268-6c26-4263-a1f7-bc09a9e3977f?standard=DMRB (accessed 14/05/2024).

National Highways (2021) CG 300 Technical Approval of Highway Structures. https://www.standardsforhighways.co.uk/tses/attachments/17dadcc6-8e01-455d-b93e-c827d280839a?inline=true (accessed 15/05/2024).

NTSB (2019) Pedestrian bridge collapse over SW 8th Street Miami, Florida Mar. 15(2018). https://www.ntsb.gov/investigations/AccidentReports/Reports/HAR1902.pdf (accessed 22/05/2024).

NTSB (2024) Contact of containership Dali with the Francis Scott Key Bridge and subsequent bridge collapse. https://www.ntsb.gov/investigations/Documents/DCA24MM031_Preliminary Report 3.pdf (accessed 14/05/2024).

OJEC (1984) Council Directive 85/3/EEC of 19 Dec. 1984 on the weights, dimensions and certain other technical characteristics of certain road vehicles. *Official Journal of the European Union* No L 2/14.

RAIB (2010) Report 02/2010: Derailment of a freight train near Stewarton. https://www.gov.uk/raib-reports/derailment-of-a-freight-train-near-stewarton-ayrshire (accessed 05/06/2024).

Sarkar (2023). 20 workers killed as crane collapses in India's Maharashtra state. *The Independent.* https://www.independent.co.uk/asia/india/maharashtra-thane-crane-collapse-bridge-death-b2385526.html (accessed 19/05/2024).

Shave J and Bennetts J (2022) Assessment of load capacity of existing highway bridges. In *Highway Bridge Management* (Cole G and Fish RJ (eds)), ICE Publishing, London, UK, pp. 73–91.

Shetty N, Chubb M, Knowles N and Halden D (1996) A risk-based framework for assessment and prioritisation of bridges. *3rd International Conference on Bridge Management*, Thomas Telford, London, UK, pp. 563–571.

Shetty N, Chubb M, Yeoell D and McFarlane R (2000) Proposed LoBEG prioritisation system for bridge strengthening, maintenance and upgrading works. *4th International Conference on Bridge Management* **4:** 744–754, Thomas Telford, London, UK.

Standards New Zealand (2004) AS/NZS 4360:2004. https://www.standards.govt.nz/shop/asnzs-43602004/ (accessed 13/05/2024).

The Guardian (2024) Baltimore bridge collapse: At least six missing as Biden laments 'terrible accident'. *The Guardian.* https://www.theguardian.com/us-news/2024/mar/26/baltimore-francis-scott-key-bridge-collapses-after-boat-collision (accessed 15/05/2024).

transportgeography.org (2024) World merchant fleet, tonnage registered per ship size, 1970-2020. The Geography of Transport Systems. https://transportgeography.org/contents/chapter5/maritime-transportation/world-registered-fleet-tonnage/ (accessed 15/05/2024).

TWF (2024) Aims and Objectives – Temporary Works Forum. https://www.twforum.org.uk/about-us/aims-and-objectives (accessed 16/05/2024).

Weiss G (2024) Thousands of US bridges are vulnerable to collapse from a single hit: NTSB. https://www.businessinsider.com/baltimore-bridge-fracture-critical-vulnerable-collapse-2024-3 (accessed 15/05/2024).

Wood S, Rees J, Fisher L, Parkes M and Evans K (2023) Hanger replacement for the Menai Suspension Bridge, Wales. *The Structural Engineer*, **101(9)**.

Richard Fish
ISBN 978-1-83608-559-1
https://doi.org/10.1108/978-1-83608-556-020251010
Emerald Publishing Limited: All rights reserved

Chapter 9
Professional responsibility and ethics

Some of the historical collapse case studies described in earlier chapters have shown how human error, either explicitly or implicitly, was a significant contributory factor that led to the failure. The spectrum of those factors ranges from an innocent mistake to potentially criminal negligence. Throw in either a lack of competence or a failure in communications (or both) and there is a complex picture of how people need to be trained, assessed and managed in order to avoid such failures. Fighting against those errors should be the ethical values not only of organisations but also of individuals. This chapter takes a step back and reflects on the importance of professionalism and ethics across the broad gambit of engineering.

9.1.   A brief excursion into philosophy

The point of this chapter is not to get embroiled in philosophical argument but rather to consider what good ethical behaviour looks like for professional engineers. Those readers who might wish to lose themselves in the subject could start with a simple introduction (Blackburn, 2021). Philosophy as a subject, however, can be helpful because the way we manage bridges is actually through people. And, as with bridges, everyone is different.

In terms of definitions, the word 'ethics' is derived from the Greek '*êthos*' which means one's personal morals or disposition. Ethics can be defined as a moral philosophy; the basis on which our personal characteristics, behaviours, values and beliefs are formed. It defines how we might react in different situations, our levels of honesty and integrity, and how we might decide what behaviours that we see in others are either acceptable or unacceptable.

At an individual level, the question of how we derive our own moral standards is one which has exercised philosophers for at least two millennia: that of nature against nurture (Pinker, 2002). In other words, to what extent is our collection of personal values hard-wired at birth and how much is influenced by interactions with the world around us. Even the 'nature' element is itself hugely dependent on our genetic inheritances and can even be attributed in part to Darwinism (Galton, 1875). The 'nurture' aspect is one that can be shaped by external influences, not least education. The extent to which nurture can modify any of those traits in our nature, however, is of course dependent on how much pre-existing resistance is already present in our psyche. Perhaps unsurprisingly, philosophers and academics continue to argue these points, although they are probably, unlikely ever to reach agreement.

9.2.   Engineering judgement, ethics and education

As has been shown in earlier chapters, engineering judgement is an important part of being a bridge manager and its application can have a significant impact on decisions and outcomes. Engineering judgement is also a personal quality and one that can be influenced by an individual's ethics or their moral compass. The question is, can engineering judgement, and indeed ethics, be taught?

Most education systems seem intent on separating early teenage students into those whose preference might be arts and humanities and those who might have a bent for maths and science. In England, they both then generally follow parallel paths through GCSE and A-levels (or their equivalents in other countries), and some remain on those trajectories to university and eventually the workplace. Admittedly, this is a generalisation and there will be some who may be the exception that proves the rule. This somewhat narrow focus, however, can mean that many engineers find it difficult when they reach senior positions to deal with situations in which there is no textbook to reach for, nor a formula to apply.

Similarly, it could be argued that they can become over-rational in their decision-making processes, especially when faced with problems that will take them outside of their comfort zone. There will be a tendency to see things only in a binary sense: only black or white, with no shade of grey in between. Such individuals, therefore, who might well also be bridge managers, can sometimes struggle with the concept of engineering judgement which will certainly make their job harder and possibly even have more serious implications.

It has been proposed that one way to teach engineering judgement to undergraduates is by subjecting them to real-life examples of engineering failures (Edmondson and Sherratt, 2022). This study exposed students in their first term, with no prior tuition in structural engineering, to some failure case studies and required them to list causes simply based on their personal judgements. The outcome was surprisingly encouraging, recognising that more than one cause was always likely and that people issues were almost always going to be highly relevant.

9.3.   Ethics and trust

Professional engineers have a duty of care towards the public at large and bridge managers specifically have that same level of duty to the travelling public. As noted in the preface, people using bridges will have such an implicit trust in engineers that they will be convinced that they are exposed to zero risk as far as the safety is concerned of the bridges that they are travelling over or under.

For a broader perspective, some case studies are given below which, with respect to public trust in our wider profession, have revealed significant shortcomings. While none of these relate specifically to civil or structural engineering, they are all highly relevant. They have been chosen as examples not only to promote the need to re-establish trust in the broader engineering profession but also to set a high bar to which engineers in all sectors should aspire.

The UK's Royal Academy of Engineering first addressed this issue in 2011 (RAEng, 2011), followed over a decade later by a joint report with the Engineering Council. The latter should be essential reading on this subject (RAEng, 2022). Another year later, the Royal Academy also commissioned an independent report (GoodCorporation, 2023) which listed six key findings.

(i)   Engineers and technicians report good ethical practice and ethical culture in engineering compared to the general UK workforce, but there are worrying signs of poor ethical standards in some parts of the profession.
(ii) There is evidence many engineers and technicians feel dissuaded from raising concerns in the workplace.
(iii) Engineers and technicians in larger firms have more support when it comes to ethics than those working in smaller firms.

(iv) Engineering firms rank the safety, health and wellbeing of workers, business integrity, and cybersecurity as the most relevant ethical risks for their organisations.

(v) Professional engineering institutions are beginning to explore ethical issues, but often in a piecemeal and unsystematic way.

(vi) A lack of integration and coordination within UK engineering creates obstacles in communication and engagement on ethics.

These findings are a little like the curate's egg: good in parts. They do show, however, that much work remains to be done in order to have the confidence that the implicit public trust can be justified.

Another useful piece of work has been produced by the Engineering Professors Council (EPC, 2024). This is an engineering ethics toolkit which, perhaps unsurprisingly, is aimed at academics who might be looking for resources to help with teaching ethics courses across all of the UK's engineering disciplines. Nonetheless, it is an interesting source of material for anyone interested in engineering ethics with a veritable rabbit warren of links to numerous articles and case studies, not only in the UK but also from around the world.

9.4.   Case studies from other engineering sectors and the built environment

The following case studies have been chosen because each presents a slightly different perspective on ethical issues. No firm judgemental conclusions are drawn, not least, as noted above, each of us will have a slightly different perception of acceptability and the reader should form their own views.

9.4.1   The Challenger space shuttle

On 28th January 1986, the NASA space shuttle *Challenger* and its crew of seven (including Christa McAuliffe, the 'teacher in space') launched from Cape Canaveral in Florida, USA. This was *Challenger's* tenth flight and the 25th of NASA's shuttle fleet. By the time it was 73 seconds into the flight, it had reached a height of 14 km when it blew up (Peake, 2023).

The investigation that followed was undertaken by a presidential commission (NASA, 1986). The cause of the disaster proved to be the failure of the O-rings, effectively elastomeric circular gaskets, in a joint on one of the solid rocket boosters. More importantly, however, the failure was attributed to a design flaw. In fact, doubts had been expressed on the efficacy of O-rings, especially in low temperatures, as early as 1977. Managers at both NASA and their shuttle contractor, Morton Thiokol (MT), were well aware of the O-ring weakness but had declined to do anything about it. The commission concluded that the tragedy was 'an accident rooted in history'.

From the ethics point of view, a lesson to be learned concerns the attempts by an MT engineer, Roger Boisjoly, to raise awareness of potential safety issues (Whitbeck, 2011). Boisjoly had first expressed concerns about O-rings when, in January 1985, he had found evidence of hot gas which had been escaping through one of the joints in sections recovered from an earlier launch. He presented his findings to a number of NASA review boards but with no effect, other than being asked to 'soften' his conclusions. His primary focus related to the O-ring material's performance in low temperatures. In the summer of 1985, Boisjoly began writing a journal to record his thoughts and actions in trying to draw attention to his concerns.

The day before the January 1986 *Challenger* launch, Boisjoly learned that the nighttime tempera-ture was forecast to fall to −8°C (18°F). He urgently contacted MT's vice-president of engineer-ing and convinced him that the launch should be postponed. This led to a teleconference with top NASA officials but ultimately the decision was made to go ahead. This was ostensibly because NASA had recently had some PR setbacks and wanted to garner increased public support for their programme. They needed a good news story which the 'teacher in space' would provide, even hop-ing for this to feature in US President Ronald Reagan's State of the Union address which was due to be given that evening.

When the presidential commission was established, Boisjoly decided that he must act as a whistleblower and gave his journal to the commission, before MT could review or even censor it. Boisjoly's honesty and courage made him a recognised advocate of engineering ethics and respon-sible decision making in the workplace. That came at some personal expense, however, as he was ostracised by MT colleagues, removed from their space programme projects and later suffered from post-traumatic stress disorder.

The whole space shuttle programme was one in which ethics was a prominent feature. Central to this were the tensions between cost, safety, quality and time. NASA's budgets, post the *Apollo* moon landings, were under pressure from the federal government, and procurement policies usu-ally meant accepting the lowest bidder in a fee competition. As was the case with *Challenger*, pressure to meet deadlines had to mean that safety or quality (or both) were going to suffer (Pinkus *et al.*, 1997).

9.4.2   The Volkswagen diesel emissions scandal

In 2015, the US Environmental Protection Agency (EPA) discovered that the German car maker, Volkswagen (VW), had deliberately programmed its turbocharged direct injection diesel engines to activate their emissions controls only during laboratory testing. This meant that the EPA's require-ments for nitrous oxide emissions were met in the tests but in production models sold worldwide, the same software controls were not employed, meaning that actual emissions were some 40 times greater than the EPA standard.

It later transpired that this had been the case since 2009 and, in the intervening six years, about 11 million cars with this defect had been sold. Although VW accepted responsibility, its CEO ini-tially denied any corporate accountability, blaming the misdemeanour on 'the terrible mistakes of a few people'. He later resigned, however, and the company had to allocate over $7 billion to recall cars (both VW and Audi) around the world to rectify the software (Aichner *et al.*, 2020).

Although the eventual responsibility was proved to be corporate, the decisions taken by the com-pany engineers, irrespective of whether they were acting alone or through decisions taken at senior management level, ask many questions from an ethical viewpoint. As well as the financial burden that this gave the company, it also caused serious reputational damage.

9.4.3   The Grenfell Tower fire

Although with no direct engineering implications, there are several ethical lessons that can be drawn from the fire that engulfed this residential tower block in London, UK, on 14th June 2017, killing 72 people.

Grenfell Tower was a typical brutalist-style 1970s tower block, just over 67 m high and with 24 storeys, owned by the local authority, The Royal Borough of Kensington and Chelsea, but since

1996 managed through an arms-length organisation. The block had undergone a major renovation including new windows and new external cladding, which had been completed in 2016. The cladding was intended to improve the insulation of the building as well as enhance its aesthetic appearance. It later transpired that the choice of cladding material, an aluminium-polyethylene composite, was driven more by cost than by its fire-resistance properties.

The fire started with an electrical fault in a refrigerator in a fourth-floor flat. The resident was awoken by his smoke detector and called the fire and rescue service at 0.54 am on the morning of 14th June. Although firefighters had entered the tower at 0.59 am, by 1.09 am the fire had penetrated the window frame of the flat and, a few minutes later, ignited the cladding. The flames quickly climbed the building. The large number of fatalities was due in part to residents following the requirements of the fire drill which was to 'stay put' in your flat and await rescue. This had been based on the presumption that a fire would be *inside* the block and, therefore, should be containable. The fact that the fire was *external,* and fuelled by the plastic in the new cladding, meant that scores of victims were either asphyxiated or burnt to death (BBC, 2018a).

The tragedy led to the UK government commissioning an independent review of the UK building regulations led by Dame Judith Hackitt (OGL, 2018). Her report described the UK building regulatory system as 'not fit for purpose'. Primary legislation has also since been enacted in the form of the Building Safety Act 2022 (legislation.gov.uk, 2022).

The government also established a public inquiry under the chairmanship of Sir Martin Moore-Bick which started in September 2017. An initial report, which considered factual evidence of what happened and when, was published in October 2019 with a final report in September 2024 (Grenfell Tower Inquiry, 2024). This report is excoriating in terms of decades of government, other public organisations and arm's-length bodies showing at best an indifference to reports and recommendations from major fires in the UK in the 1990s.

With specific regard to the cladding, the large number of organisations involved in the management, procurement, delivery and implementation of the refurbishment of the building created such obfuscation that tracing responsibility and accountability has been very difficult. The inquiry's report, however, leaves no party blameless, with some being described as dishonest or incompetent, as well as taking a 'casual approach to contractual relations' (Grenfell Tower Inquiry, 2024, Phase 2 Report, pp. 18). As well as the poor ethical behaviours that have been exposed by the inquiry, it is highly likely that criminal investigations will follow, considering the actions of individuals and organisations who could face potentially significant charges.

9.4.4   Boeing 737 MAX 8 airliner

The American company Boeing is the world's biggest manufacturer of commercial aircraft. In 2016, the fourth generation of its popular 737 series, the MAX 8, made its maiden flight. In October 2018, a 737 MAX 8, Lion Air flight 610, crashed into the sea 13 minutes after leaving Tangerang, Indonesia, killing all 189 people on board (BBC, 2018b). A few months later, in March 2019, another 737 MAX 8, Ethiopian Airlines flight 302, crashed only six minutes after taking off from Addis Ababa, claiming 157 lives (BBC, 2019). As a result of these two incidents, all 737 MAX series aircraft were grounded.

Investigations into both crashes revealed that the aircraft's computer systems had activated the 'Maneuvering (sic) Characteristics Augmentation System' (MCAS – a new automatic flight control feature on the MAX variant) but based on erroneous flight data. This had caused both aircraft to pitch into a dive and, even after the pilots' attempts to disable it, the MCAS continued to function and ultimately led to the tragedies.

The investigations eventually led to the revelation that Boeing had declared that MCAS was simply an addition to existing systems rather than something completely new, and hence it was not given prominence in the aircraft manuals, nor for pilot training in simulators. It transpired that there had been a conscious decision *not* to inform the US Federal Aviation Authority (FAA) as this would have required additional certification with a consequential delay in the aircraft coming into service.

The case against Boeing was considered by the US Department of Justice (DOJ) and a settlement was reached in which Boeing would pay fines and compensation totalling $2.5 billion (DOJ, 2021). Some extracts from the DOJ report are particularly damming.

> *'The tragic crashes ... exposed fraudulent and deceptive conduct by employees...'*

> *'Boeing's employees chose the path of profit over candor (sic) by concealing material information from the FAA...'*

> *'The misleading statements, half-truths, and omissions communicated by Boeing employees to the FAA impeded the government's ability to ensure the safety of the flying public.'*

The lack of ethical decision making as found by the FAA and as exemplified in the above quotes is frightening. Chapter 1 made a comparison of risk to the public between airliners and bridges. This episode demonstrates that the management and leadership in the former must be based on ethics; and the same principles must also apply to bridges.

Of course, as well as the immediate financial penalty, Boeing suffered from not only a significant loss of orders during the 20 months that the 737 MAX was grounded but also immeasurable and long-lasting reputational damage.

9.4.5   Professional engineers and environmental protests

Early on 18th October 2022, two protesters from the UK campaign group Just Stop Oil climbed stay cables on the Queen Elizabeth II Dartford Crossing bridge carrying the southbound M25 motorway over the River Thames to the east of London. Having equipped themselves with hammocks, the pair remained at the top of the bridge's north tower for 37 hours. For safety reasons, the bridge had to be closed during this period, causing widespread traffic congestion as well as a significant social and economic impact (Johnson, 2022). Both were jailed for the offence of causing a public nuisance (BBC, 2023).

One of the pair, a New Zealand civil engineer but working in London, stated

> *'As a professional civil engineer, each year as I renew my registration, I commit to acting within our code of ethics, which requires me to safeguard human life and welfare and the environment.'*

The big question here is to ask whether he was justified in taking action in this way as part of what he saw as his professional duty?

Probably unrelated to this specific incident, in May 2024, Engineering New Zealand (ENZ – the body covering all engineering disciplines) issued a practice note entitled 'Climate Action – The

Role of the Engineer' (ENZ, 2024). This gives guidance on legitimate action, such as embedding sustainability and carbon reduction, which all engineers should be taking in the course of their work. It also highlights the responsibility of its members to uphold their Code of Ethical Conduct.

9.5. Advice on ethical conduct

As noted in Section 9.3, the 2023 report (GoodCorporation, 2023) for the Royal Academy of Engineering (RAEng), Ethics in the Engineering Profession, included the finding that 'professional engineering institutions are beginning to explore ethical issues, but often in a piecemeal and unsystematic way'. Although this statement seems to be both glass-half-full and glass-half-empty, there has been some good progress not only in raising the profile of ethics within professional engineering institutions but also making individual members more aware of their ethical responsibilities (Rose, 2024).

Back in 2005, the UK Engineering Council (together with the RAEng) produced an initial statement of ethical principles which was updated in 2017 (Engineering Council, 2017). This statement has four main areas of focus

(i)   honesty and integrity
(ii) respect for life, law, the environment and public good
(iii) accuracy and rigour
(iv) leadership and communication.

The statement was adopted by the Institution of Structural Engineers and developed as a business practice note entitled 'Ethics' in the Institution's monthly journal (Entwhistle, 2017).

The Institution of Civil Engineers reconstituted its Ethics Committee in its governance in 2021, setting itself five ethical challenges (ICE, 2024)

- climate change
- corruption
- fairness, inclusivity and respect
- project governance and risk
- ICE code of ethics.

The last of these has now been approved by the ICE Council and relaunched as the 'ICE advice on ethical conduct' (ICE, 2023a), updating versions published in 2004 and 2012, and is intended to be a sister document for the institution's Code of Professional Conduct (Luck, 2023).

The primary purpose of each of these documents, and their equivalents in other countries, is to help guide engineers in their decision-making processes; to be a litmus test to confirm or otherwise their ethical instincts.

9.6. Ethical actions – whistleblowing

As with the case of Roger Boisjoly noted in Section 9.4.1, a whistleblower is someone who reports something that is happening in an organisation which is either illegal, unsafe or fraudulent, not for personal gain but for the common good. The reporting is usually to a senior manager within the same organisation as the whistleblower, but can also be to an external agency such as the police or the media. While it may be the whistleblower's preference to remain anonymous, that is not always possible. Whistleblowing is not without risk, especially if not anonymous; not only on a personal level with potential mental health and wellbeing issues but also with the impact it can have on friends and family. The decision on whether to whistleblow is very much one of ethics and an individual's personal values.

In the UK, there are legal safeguards in place to protect whistleblowers in most workplace situations (ACAS, 2023). This principally comes under the Public Interest Disclosure Act 1998 (legislation.gov.uk, 1998) but most large organisations will probably have a whistleblowing policy which should be readily accessible.

9.7.   Ethical actions – confidential reporting

In the civil and structural engineering sectors, and specifically with respect to bridges, the facility exists to effectively whistleblow, with complete anonymity, to CROSS-UK, or sister CROSS organisations in other countries. Under a previous incarnation the acronym stood for Confidential Reporting on Structural Safety. Although that has now changed, the principle remains the same and the CROSS-UK website contains a significant list of anonymised confidential reports, all having been submitted in the hope that we will learn from the experiences of others (CROSS-UK, 2024a).

9.8.   Ethical challenges – corruption

One of the biggest ethical challenges facing every sector is the threat of corruption, either at a relatively low level with employees being tempted by bribes or inducements or at a more corporate level with potentially fraudulent activity taking place. The latter could be either with respect to a misrepresentation of accounts or by malfeasance in the process of winning contracts ahead of a competitor. While fraud may be at a relatively low level in the UK, temptations will arise in some other countries where there might be a culture which has much lower acceptable levels of probity. Unfortunately, the construction sector is one of the most fertile when it comes to these matters.

In the UK, the organisation which seeks to address this issue is the Anti Corruption Forum, founded in 2004 (anticorruptionforum.org.uk, 2024). This is a body which has representation from over 1000 UK companies and some 350 000 professionals. Although UK orientated, the Forum is well aware that a significant challenge for UK companies, who are aiming to secure work overseas in countries where bribery is almost part of their national identity, is to be able to maintain their integrity. This is not only a commercial issue but also one of ethics in which the stand taken by company directors, shareholders and staff will give a strong indication of their ethical culture.

9.9.   Ethical challenges – competence

The need to enhance competence at all levels of every sector of engineering is one of the biggest challenges facing our profession, and probably none more so than in civil and structural engineering. In the UK and in the context of both publicly funded capital projects and in the revenue funding of asset management, most of this century has been earmarked by government demanding the need for efficiency, value for money, reducing waste, and various initiatives such as best value and right first time. Additional pressures arise from the growth of a litigious culture and a focus on professional indemnity insurance in the event that mistakes are made. Although all very noble and necessary, what has been the outcome from the law of unintended consequences?

For example, in 2018, the UK's Office for National Statistics (ONS) estimated that the contribution of the construction sector to the national economy was £100 billion. It also reported that about 10 per cent of this figure – £10 billion – was soaked up in errors and in litigation. Much of the

10 per cent is attributable to a lack of experienced and suitably qualified engineers and an over-reliance on management systems (Burrell, 2019).

The underlying issue here is one of competence, at every level. Part of the reason that there is a perceived decline in professional competence lies in commercial pressures to cut costs, to be more efficient in order to secure work. This is the progressive extrapolation of procurement initiatives and fee competition. In order to win work, fees must be minimised, and hence junior (i.e. cheaper and less experienced) members of staff are assigned to the task. Staff who may not yet be professionally qualified could be thrust into having to make decisions which they are unqualified to make.

More often than not, the basis of their decisions comes directly from the results from, and an over-reliance on, structural analysis software. There is either no time nor, occasionally, even the right skill set to carry out a simple hand calculation just to have confidence that the results are in the right area. No matter how sophisticated the software and how well proven is its efficacy, the well-known adage of rubbish in rubbish out will apply. This implicit 'trust in the machine' must be one of the biggest concerns of twenty-first-century engineering. A recent example here is that cited in a CROSS-UK Safety Report concerning the design of a relatively simple retaining wall using bespoke software (CROSS-UK, 2024b).

Measuring and testing competence should be a vital issue for all organisations. While on-the-job training and mentoring are important elements, so too is ensuring that all professional engineers are keeping up with their own continuing professional development (CPD). Although there is an ongoing debate on whether CPD should be mandatory, both the Institution of Civil Engineers (ICE, 2023b) and the Institution of Structural Engineers (IStructE, 2024) have mandated the *recording* of CPD. Both carry out annual audits on a sample of members to help ensure compliance. Not all staff in this discipline of engineering, however, are members of professional institutions and more work is needed to drive up competence, particularly as noted in Chapter 7 regarding bridge inspectors (LANTRA, 2016).

Irrespective of where anyone sits in any organisational hierarchy, there should be a personal ethical dynamic to only undertake work for which they have the confidence to acknowledge their limitations. Furthermore, as more responsibility is accepted, they need to know when those limits are going to be exceeded. Competence should be allowed to grow with experience, education and training but it should also be regularly tested throughout our careers.

9.10. An Archimedic oath?

In the UK, members of the medical profession swear an oath of ethics. This is the Hippocratic oath, traditionally attributed to the Greek physician, Hippocrates of Kos, often referred to as the 'father of medicine'.

Is there a case for engineers to swear a similar oath of ethics? Perhaps our closest equivalent of Hippocrates might be Archimedes: a reasonable fit if we choose the derivation of 'engineer' as coming from 'ingenuity'. This book is not the place to consider how such an oath might be drafted but if it would help us work to higher ethical standards, then why not develop the idea?

A country whose engineers can, if they choose, swear an oath is Canada. Those that do so are also given an iron ring to be worn on the little finger of their dominant hand where it might tap as they write or draw (or now type or use a mouse) to remind them of their professional and ethical responsibilities. This all takes place at a private ceremony known as 'the Obligation of an Engineer' or 'the ritual of the calling of an engineer', first instigated in 1925 (Petroski, 2012). The custom is endorsed by, but independent from, the Engineering Institute of Canada. In 1970, the USA adopted a similar practice.

There is an apocryphal story that the rings are made from material recovered from the wreckage of the Quebec bridge which collapsed under construction in 1907, killing 75 workers. Although this was a steel (and therefore not iron!) bridge, the ethical lessons that should be learned from the failure still resonate today: that the designer failed to calculate the dead load correctly, the resident engineer ignored warnings that the steel was overstressed, and the chief consulting engineer had delegated decision-making responsibilities to young and inexperienced engineers.

'The Obligation of an Engineer' is administered by the Corporation of the Seven Wardens, a reference to the seven past presidents of the Engineering Institute of Canada who supported the initiative when first mooted in 1922 (Corporation of the Seven Wardens, 2024). The Corporation's very worthy mission and vision statements are examples which most countries would do well to emulate.

MISSION: *'To obligate Canada's engineering community to ethical conduct'.*

VISION: *'Lifelong ethical conduct by Canada's Engineers'.*

9.11. Conclusion

Although it could be argued that this chapter may have gone off on a tangent to the central theme of the book, the deviation is an important one. Almost all the examples of bridge collapses or failures discussed in earlier chapters, from the Ponte Milvio in AD 312 to the Francis Scott Key Bridge in 2024, as well as those case studies from other engineering sectors in this chapter, and that of the Quebec bridge, have people issues as a contributory cause. And central to most of those people issues has been an individual or an organisation, either knowingly or unwittingly, acting unethically. If such events are to be prevented in future, raising the profile of ethics in education, engineering institutions and the workplace must play a big part in achieving that goal.

REFERENCES

ACAS (2023) The law – Whistleblowing at work. https://www.acas.org.uk/whistleblowing-at-work (accessed 28/05/2024).

Aichner T, Coletti P, Jacob F and Wilken R (2020) Did the Volkswagen Emissions Scandal Harm the 'Made in Germany' Image? A Cross-Cultural, Cross-Products, Cross-Time Study. *Corporate Reputation Review.* https://link.springer.com/article/10.1057/s41299-020-00101-5 (accessed 26/05/2024).

anticorruptionforum.org.uk (2024) Anti Corruption Forum – Preventing Malfeasance. https://www.anticorruptionforum.org.uk/ (accessed 28/05/2024).

BBC (2018a) Concerns raised about Grenfell Tower 'for years'. https://www.bbc.com/news/uk-england-london-40271723 (accessed 28/05/2024).

BBC (2018b) Lion Air crash: Boeing 737 plane crashes in sea off Jakarta. https://www.bbc.com/news/world-asia-46014463 (accessed 27/05/2024).

BBC (2019) Ethiopian Airlines: 'No survivors' on crashed Boeing 737. https://www.bbc.com/news/world-africa-47513508 (accessed 27/05/2024).

BBC (2023) Just Stop Oil: Dartford Crossing protesters jailed. https://www.bbc.com/news/uk-england-essex-65263650 (accessed 28/05/2024).

Blackburn S (2021) *ETHICS: A Very Short Introduction*, 2nd edn. Oxford University Press, Oxford, UK.

Burrell P (2019) What lies behind structural errors and failures within the UK. *Proceedings of the Institution of Civil Engineers. Forensic Engineering* (**4**): 171–178.

Corporation of the Seven Wardens (2024) Home. https://ironring.ca/home-en/ (accessed 31/05/2024).

CROSS-UK (2024a) Full category index. https://www.cross-safety.org/uk/safety-information/full-category-index (accessed 28/05/2024).

CROSS-UK (2024b) Unqualified engineer's unsafe computer aided design of a retaining wall. https://www.cross-safety.org/uk/safety-information/cross-safety-report/unqualified-engineers-unsafe-computer-aided-design-1210 (accessed 31/05/2024).

DOJ (2021). Office of Public Affairs: Boeing charged with 737 Max fraud conspiracy and agrees to pay over $2.5 billion. United States Department of Justice. https://www.justice.gov/opa/pr/boeing-charged-737-max-fraud-conspiracy-and-agrees-pay-over-25-billion (accessed 27/05/2024).

Edmondson V and Sherratt F (2022) Engineering judgement in undergraduate structural design education: Enhancing learning with failure case studies. *European Journal of Engineering Education* **47**(**4**): 577–590.

Engineering Council (2017) Engineering Council. https://www.engc.org.uk/professional-ethics (accessed 28/05/2024).

Entwhistle J (2017) Business practice note. No9: Ethics. *The Structural Engineer* **95** (**9**): 20–21.

ENZ (2024) Practice Note 32, Climate Action - the role of the engineer. https://d2rjvl4n5h2b61.cloudfront.net/media/documents/PN32_ClimateAction.pdf (accessed 28/05/2024).

EPC (2024) Engineering Ethics Toolkit – Engineering Professors Council. https://epc.ac.uk/resources/toolkit/ethics-toolkit/ (accessed 26/05/2024).

Galton F (1875) On men of science, their nature and their nurture. *Proceedings of the Royal Institution of Great Britain* **7**: 227–236.

GoodCorporation (2023) *Ethics in the Engineering Profession*. GoodCorporation Ltd., London. UK.

Grenfell Tower Inquiry (2024) Ministry of Housing, Communities and Local Government, UK. Homepage, Grenfell Tower Inquiry. https://www.grenfelltowerinquiry.org.uk/ (accessed 05/09/2024).

ICE (2023a) ICE advice on professional conduct. https://www.ice.org.uk/media/l3ajpjlr/advice-on-ethical-conduct-ice-ethics-committee-december-2023.pdf (accessed 28/05/2024).

ICE (2023b) Continuing Professional Development Guidance. https://view.officeapps.live.com/op/view.aspx?src=https%3A%2F%2Fwww.ice.org.uk%2Fmedia%2Frjknbo5g%2Fcontinuing-professional-development-guidance.docx&wdOrigin=BROWSELINK (accessed 31/05/2024).

ICE (2024) Ethics committee, Institution of Civil Engineers (ICE). https://www.ice.org.uk/about-us/who-we-are/ethics-committee#more-in-this-section (accessed 28/05/2024).

IStructE (2024) Continuing Professional Development (CPD) – The Institution of Structural Engineers. https://www.istructe.org/training-and-development/cpd/ (accessed 31/05/2024).

Johnson T (2022) Civil engineer protesting on Dartford Crossing driven by professional duty to protect environment. *New Civil Engineer*. https://www.newcivilengineer.com/latest/civil-engineer-protesting-on-dartford-crossing-driven-by-professional-duty-to-protect-environment-18-10-2022/ (accessed 28/05/2024).

LANTRA (2016) *National Sector Scheme 31 for the Bridge Inspector Certification Scheme*. https://www.lantra.co.uk/national-highway-sector-schemes-nhss/bridge-inspectors (accessed 26/04/2024).

legislation.gov.uk (1998) Public Interest Disclosure Act 1998. https://www.legislation.gov.uk/ukpga/1998/23/contents (accessed 28/05/2024).

legislation.gov.uk (2022) Building Safety Act 2022. https://www.legislation.gov.uk/ukpga/2022/30/contents/enacted (accessed 28/05/2024).

Luck L (2023) Do we always know how to make ethical decisions in the workplace? Institution of Civil Engineers (ICE). https://www.ice.org.uk/news-insight/news-and-blogs/ice-blogs/the-civil-engineer-blog/do-we-know-how-to-make-ethical-decisions-at-work (accessed 28/05/2024).

NASA (1986) Rogers Commission Report. https://sma.nasa.gov/SignificantIncidents/assets/rogers_commission_report.pdf (accessed 27/05/2024).

OGL (2018) Building a Safer Future: Final Report. https://assets.publishing.service.gov.uk/media/5afc50c840f0b622e4844ab4/Building_a_Safer_Future_-_web.pdf (accessed 28/05/2024).

Peake T (2023) *SPACE: The Human Story*. Century Publishing, London, UK.

Petroski H (2012) *To Forgive Design: Understanding Failure*. Harvard University Press, Cambridge, USA.

Pinker S (2002) *The Blank Slate: The Modern Denial of Human Nature*. Penguin Books, London, UK.

Pinkus R, Schuman L, Hummon N and Wolfe H (1997) *Engineering Ethics: Balancing Cost, Schedule and Risk; Lessons Learned from the Space Shuttle*. Cambridge University Press, New York, USA.

RAEng (2011) *Engineering Ethics in Practice: A Guide for Engineers*. Royal Academy of Engineering, London, UK.

RAEng (2022) *Engineering Ethics: Maintaining Society's Trust in the Engineering Profession*. A report by the Engineering Ethics Reference Group established by the Royal Academy of Engineering and the Engineering Council, Royal Academy of Engineering, London, UK.

Rose C (2024) Does Ethics really play a role in civil engineering? *New Civil Engineer*. https://www.newcivilengineer.com/opinion/does-ethics-really-play-a-role-in-civil-engineering-28-02-2024/ (accessed 30/05/2024).

Whitbeck C (2011) *Ethics in Engineering Practice and Research*, 2nd edn. Cambridge University Press, New York, USA.

emerald PUBLISHING   ice

Richard Fish
ISBN 978-1-83608-559-1
https://doi.org/10.1108/978-1-83608-556-020251011

Chapter 10
Climate change implications

Earlier chapters have reflected on the many challenges facing bridge managers trying to reduce the risk of failures in an ageing bridge stock. Such pressures are certainly going to increase if the metrics around which a bridge was originally designed start to change. While the most obvious of these may be design live load (and we have seen in Chapter 8 the significance of such changes in the UK in the 1990s), possibly an even greater threat comes from the additional environmental demands on a structure caused, either directly or indirectly, by the effects of climate change.

10.1. Background and context

Although not immediately obvious from the actions taken by politicians around the world during most of this century, the widely accepted view is that the current climate crisis poses the greatest ever existential threat to life on our planet. Both scientific and circumstantial evidence that things are changing at an ever-increasing pace now seems incontrovertible. And even any remaining sceptics who believe that the dramatic increase in carbon emissions since the Industrial Revolution is merely a coincidental factor, and that the earth is warming as part of a natural cycle, cannot deny that change is happening.

Fortunately, there are many organisations and individuals who have been doing everything they can to draw attention to the impending crisis and to press for action to be taken. One of the first leading politicians to do so in a very demonstrative way was the former US Vice President and presidential candidate in 2000, Al Gore, with his 2006 film *An Inconvenient Truth* (Guggenheim, 2006). A few years later, US President Barack Obama addressed the United Nations in a speech which included this famous quote (Obama, 2014).

'We are the first generation to feel the effect of climate change and the last generation who can do something about it.'

The United Nations sometimes seems to have been the only organisation of any influence to have taken climate change seriously and to try to push national governments to take action. Under the auspices of the United Nations Framework Convention on Climate Change (UNFCCC), it has established annual conferences of UNFCCC 'parties', each now known as a Conference of the Parties (COP). The first COP was held in Berlin, Germany, in 1995, with COP29 which was held in Baku, Azerbaijan, in November 2024.

Although COPs seem to be as much about royalty and politicians being given a platform for speeches, there is always a great deal of work going on behind the scenes to try to arrive at a form of words that will be acceptable to all countries, albeit mostly measured by the likely support (or otherwise) from home audiences. COP21 in Paris, France, in 2015, produced a legally binding international treaty which became known as the Paris Agreement, which came into force in 2016 (United Nations, 2016). The aim was to keep global temperature rises to less than $2\,°C$ when compared to pre-industrial levels, but with a preferred target of a rise of just $1.5\,°C$.

As early as 1988, the United Nations also established the Intergovernmental Panel on Climate Change (IPCC) with the aim of providing policymakers around the world with regular scientific assessments on the latest knowledge and evidence on climate change. Its sixth report, issued in 2023 (IPCC, 2023), predicts that the Paris Agreement's 1.5 °C target will be reached in the 'near term'. It also includes commentary on climate change impacts and our preparedness in terms of mitigation, adaptation and vulnerability.

As well as the specific issue of climate change, the United Nations has also addressed the wider implications for humankind through its 17 Sustainable Development Goals (SDGs) (United Nations, 2015) as shown in Figure 10.1. While it is only SDG 13 'Climate Action' that is the specific goal on this topic, many others can also be related to consequential impacts, demonstrating their interdependency.

Figure 10.1 United Nations Sustainable Development Goals (courtesy of the United Nations, https://www.un.org/sustainabledevelopment)

In the UK, the 2008 Climate Change Act (legislation.gov.uk, 2008) mandates a climate change risk assessment to be published every five years, with the last being in 2022 (CCRA, 2022). The eight headline risks from this assessment cite ongoing issues which might appear to be contrary to the UK aim of meeting the SDGs.

- risks to the viability and diversity of terrestrial and freshwater habitats and species from multiple hazards
- risks to soil health from increased flooding and drought
- risks to natural carbon stores and sequestration from multiple hazards
- risks to crops, livestock and commercial trees from multiple climate hazards

- risks to supply of food, goods and vital services due to climate-related collapse of supply chains and distribution networks
- risks to people and the economy from climate-related failure of the power system
- risks to human health, wellbeing and productivity from increased exposure to heat in homes and other buildings
- multiple risks to the UK from climate change impacts overseas.

Returning to the need to address bridge failures in this context, the fifth risk in the above list seems the most germane: the 'risks…due to climate-related collapse of…distribution networks'.

Interestingly, a link has been made between the previous chapter on ethics and the approach that engineers should be taking to address climate change (Francis, 2021), suggesting that it is our collective ethical and moral duty to do all in our power to address the challenges of the climate crisis.

10.2. What to expect

Globally, 2023 was the warmest year on record (to date) with a global average temperature of 1.46 °C above the pre-industrial period. It was also the tenth successive year that had been greater than, or equal to, 1.0 °C above the same baseline (Met Office, 2023). As well as these more obvious increases in global temperatures, the UK's Meteorological Office also produces many other climate projections (Met Office, 2022). Another valuable resource is the Climate Action Tracker (CAT), covering 39 countries and the European Union which, between them, are responsible for 85% of global greenhouse gas emissions (CAT, 2024). Based in Berlin, Germany, CAT is a collaboration between two organisations: Climate Analytics and the New Climate Institute. Figure 10.2 shows the CAT thermometer as of December 2023.

The rate of change, however, is by no means linear. The US National Oceanic and Atmospheric Administration (NOAA, 2024) hosts a specific website with a climate dashboard (climate.gov, 2024). Some of its key readings (as of June 2024) are given below.

- The heating influence of all human-produced GHGs* was 49% higher in 2022 than it was in 1990.
- Since 1979, the extent of ice covering the Arctic Ocean at the end of summer has shrunk by more than 40 per cent.
- Atmospheric carbon dioxide has risen by more than 50 per cent since people began burning fossil fuels for energy.
- The oceans are storing 91 per cent of the excess heat from global warming, causing sea level rise, ice shelf retreat, and stress on marine life.

Among the most problematic global consequences of these effects are the melting of the polar ice caps, sea level rise, the thawing of the Arctic tundra (with a consequent release of stored methane), more intense and more frequent storm events (hurricanes, typhoons, cyclones, etc.), and an increase in forest or wild fires (not only pumping more carbon into the atmosphere but also, at the same time, reducing the world's forestation needed to reabsorb carbon dioxide). All of these serve to accelerate the rate of global warming, making predictions that were originally long term now much nearer to the medium term (Dora and Ferranti, 2024).

---

*Greenhouse Gases

Figure 10.2 The CAT thermometer (courtesy of CAT)

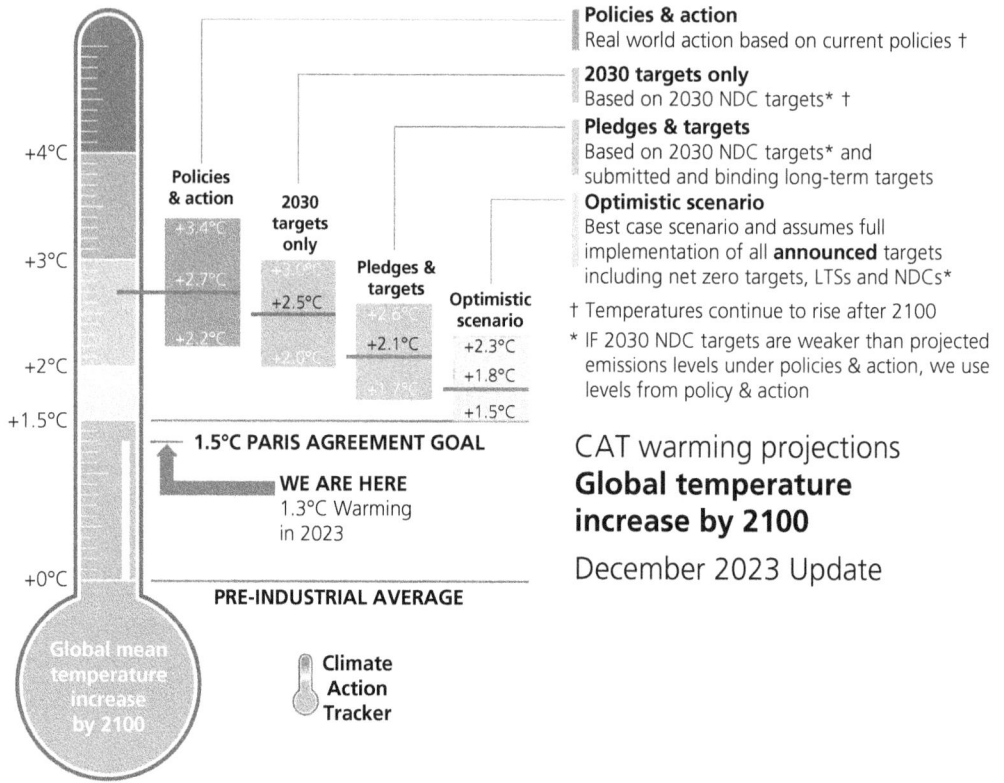

Policies & action
Real world action based on current policies †

2030 targets only
Based on 2030 NDC targets* †

Pledges & targets
Based on 2030 NDC targets* and
submitted and binding long-term targets

Optimistic scenario
Best case scenario and assumes full
implementation of all **announced** targets
including net zero targets, LTSs and NDCs*

† Temperatures continue to rise after 2100

* IF 2030 NDC targets are weaker than projected
emissions levels under policies & action, we use
levels from policy & action

CAT warming projections
**Global temperature
increase by 2100**
December 2023 Update

Policies & action +3.4°C +2.7°C +2.2°C

2030 targets only +2.5°C +2.0°C

Pledges & targets +2.1°C +1.7°C

Optimistic scenario +2.3°C +1.8°C +1.5°C

1.5°C PARIS AGREEMENT GOAL

WE ARE HERE
1.3°C Warming
in 2023

PRE-INDUSTRIAL AVERAGE

Global mean temperature increase by 2100

Climate Action Tracker

As the themes of this book are to address, and to hope to prevent, bridge failures, as with everything else, the focus needs to be twofold: both mitigation and adaptation. Mitigation (i.e. reducing carbon emissions) is hugely important not only for new bridge construction but also for the maintenance of existing. Construction, and specifically concrete and cement production, is the second-largest carbon-intensive industry in the world and is continuing to rise, mostly due to China and its recent and ongoing infrastructure expansion. In fact, it has been claimed that China poured more concrete in three consecutive years in the 2010s than the USA had used in the whole of the twentieth century (Wallace-Wells, 2019). On the basis that every little helps, it is important that every opportunity and every initiative is taken to reduce carbon in bridge operations, as well as in construction and maintenance. Our biggest problem in hoping to reduce carbon emissions is that there is a decades-long lag between cause and effect (Mulhern, 2020). In other words, even if our greenhouse gas emissions could be switched off entirely, global temperatures would continue to rise for at least another 20 years.

It is adaptation, therefore, which must be the primary focus of bridge managers' attentions if the rate of bridge collapses due to changing climatic effects is not going to increase in the next few decades.

Some specific climate actions on our structures which will need to be addressed are covered in the following sections.

10.3. Rainfall

As noted above, one of the impacts of climate change is an ever-increasing frequency and severity of storm events, often leading to highly localised and more intense rainfall than might have been the norm even at the beginning of this century.

Although not specifically bridge related, one of the most dramatic and deadly events in recent times occurred in the Mediterranean Sea in 2023. This was Medicane Daniel (Medicane being derived from Mediterranean Hurricane) which had begun as a series of extreme rainstorms in early September which caused devastating floods in Greece, Bulgaria and Turkey. The individual storms were then transformed into a surface cyclone which eventually arrived in Libya in north Africa, with catastrophic consequences. Intense rainfall landed on the Akhḍar Mountains in the north of the country on the night of 11th/12th September. As the floods swept down the valley, two dams burst and the deluge hit the city of Derna, reportedly leading to between 5000 and 15 000 fatalities (Hewson et al., 2023). Although the scale of this disaster far outweighs questions about whether individual bridges survived or collapsed, it is the intensity of the event which should serve to remind us of the threats of similar inundations to our bridge stocks.

A not dissimilar event hit the Valencia region of eastern Spain in late October 2024, leading to over 200 fatalities. One weather station recorded 491 l/m2 of rain in just eight hours, equivalent to the amount normally expected in one year (World Meteorological Organization, 2024). Again, in disasters of this scale, whether bridges collapsed or not is not a significant issue.

The effects of excessive rainfall on a bridge can range from the overwhelming of the various drainage systems (which in themselves can lead either to element failure or at least to the need for reactive maintenance interventions) through to total collapse. The latter will have been by way of the mechanisms of foundation scour, hydrostatic pressure on the upstream face of a bridge, or for masonry arches, the longer-term effects of fill saturation over the arch barrel or between wing walls. Each of these potential collapse mechanisms is discussed in more detail below.

10.3.1 Scour

Scour, or washout as it is still occasionally known, has been the principal cause of bridge collapses (certainly in the UK but doubtless the same elsewhere) probably throughout history but at least in the twenty-first century. Scour is the process of hydraulic action which causes material from the bed of the watercourse to be removed under certain flow conditions. When the bed material is near or under a bridge foundation then a full or partial collapse is a likely outcome. The various types of scour, and best practice in terms of scour management, have been comprehensively covered elsewhere (Dentith, 2022).

A variation on this theme, however, is one in which the impacts of climate change are going to add an additional level of jeopardy. This relates to the phenomenon of scour exacerbation caused by debris being carried downstream as part of the higher water levels and velocities (Ebrahimi et al., 2020). This mostly relates to debris already on the banks of rivers which may have been part of a routine vegetation clearance programme or from larger trees uprooted by the torrent of water. The likelihood of the latter is also a factor of climate change; hot conditions in times of drought will potentially weaken root stability. When this is followed by torrential rainfall, more trees will be uprooted and carried downstream until they meet an obstruction – such as a bridge – with the likelihood of causing a dramatic change in flow conditions in the vicinity of foundations.

A not insignificant problem in managing scour is the process of monitoring scour hole formation and development. In the high flow velocities that occur during a flood event, any type of inspection is exceptionally difficult, if not impossible. A further problem is that scour holes can form during flood conditions but can then refill, at least partially, with waterborne sediment being carried by the flow at high velocities being deposited into those voids as flows subside. An inspection after an event, therefore, may not be able to reveal the full picture of how close a bridge may have been to a scour-induced collapse.

In the UK, a new standard covering the management of scour has recently been introduced (National Highways, 2024). This not only describes the inspection and assessment processes but also prescribes a risk management approach to the vulnerability of scour-susceptible structures, including other hydraulic actions such as those caused by debris accumulation.

10.3.2   Differential lateral hydrostatic loading

Irrespective of the added complications of local effects, a difference in water level between the upstream and downstream elevations of a bridge will lead to both a resultant horizontal load and, potentially for multispan arch bridges with a significant area of spandrel wall, an overturning action on the bridge as a whole. While in normal conditions, such load effects will be relatively modest in the context of the mass of the global structure, they can also be significantly enhanced by a similar entrapment of debris against the upstream elevation.

This, however, is not just a case of statics. The additional dynamic loading, not least with large pieces of waterborne debris striking the bridge, adding to the already turbulent flow conditions, can mean that the overall effect is significantly magnified.

While catastrophic failure from this scenario, however, is unlikely, it is more probable that the turbulent flows around the bridge will generate additional scour below or adjacent to the founda-tions. A recent example of such a collapse concerned the Hassanabad Bridge in Pakistan. This was a multispan concrete arch bridge with open spandrels crossing the Hunza Valley. Its demise was primarily associated with extreme temperatures which had rapidly accelerated the melting of glaciers in the mountains above the valley. The unprecedented flows generated by the volume of flood water led to the complete failure of the bridge in May 2022 (Chatta, 2022).

10.3.3   Fill saturation

An additional concern with masonry arch bridges in major flood events is the effect of fill satura-tion when a structure is submerged for a significant period of time (Hulet *et al.*, 2006).

The effects here are twofold. Firstly, with a submerged arch, buoyancy factors need to be taken into account. Applying Archimedes' principle, that the buoyancy force on a submerged object is equal to the weight of water displaced, suggests that the unit weights of those submerged components can be significantly reduced. As the load-carrying capacity of an arch is based on it having no tensile properties, and, therefore, effectively reliant on the composite action of the masonry and backfill, reductions in unit weights can also reduce load-carrying capacity in the short-term.

The second point is that saturated backfill may not drain at the same rate as the surrounding water when flood levels subside. Although this may be a lesser concern with a significant area of open joints in the arch barrel and spandrels, for those elements that have sound pointing, as will usually be the case with brick arches, the saturated fill could exert an additional (albeit short-term) hydrostatic pressure on spandrel and wing walls as well as imposing an additional load on the arch extrados. The former could lead to spandrel movement over the face voussoirs of the arch or even failure.

10.4. Temperature

Perhaps the most obvious consequence of climate change is the rise in temperature. Although it is important to distinguish between climate and weather, it is the latter which will have an immediate impact on bridge behaviour. Almost irrespective of global location, the likelihood of a prolonged spell of unusually high summer temperatures is becoming the norm. Similarly, and somewhat counterintuitively, prolonged cold snaps are also not unlikely.

Although temperature effects in isolation are unlikely to be the only contributory factor to a bridge failure, when combined with others they can be of significance. As those effects will be exacerbated by climate change, it is important that our understanding of thermal actions is kept updated (Nepomuceno *et al.*, 2022).

The three main aspects of temperature in bridge decks are covered below.

10.4.1 Global temperature

Regardless of structural form and material, the expansion and contraction of a bridge deck is directly proportional to the change in temperature of the bridge itself. As will be well known, this phenomenon is generally seasonal as there is a significant lag between bridge temperature and the ambient air temperature.

Of specific interest in this regard is the need to ensure that such global thermal movements can be accommodated. For what might be considered as a conventional bridge, movement will take place at at least one (although more often than not, several) expansion joint(s). As deck temperature and hence movement increases, it is important that the expansion joints are able to continue to accommodate the full movement range.

Movement will also be taking place at bridge bearings, which will also need to be assessed to ensure that they are acting as had been originally intended – for example, as a bearing top plate slides over its lower counterpart, there will be a change (albeit very small) in the eccentricity of loading onto the substructure below, the consequences of which will need to be taken into account in ensuring that there are no other structural implications.

Many more recent bridge designs (with a total length of up to 60 m) have eliminated the need for both expansion joints and bearings. Intended to reduce the need for maintenance interventions over their design lives, these are integral bridges which have been designed to absorb thermal strains in the deck itself and/or in the fill behind the abutments (England *et al.*, 2000). As the decades go by and temperatures continue to increase, however, the performance of integral bridges may also need to be the subject of increased attention and vigilance.

10.4.2 Differential temperature

Also linked to seasonal variations, differential temperature refers to the latent summer heat within a bridge deck which, as the winter approaches, remains relatively warm within the core of the section as the extremities, the soffit and the carriageway surfacing, begin to cool. Similarly, the opposite effect will occur after the winter when a relatively cold deck section starts to warm in the summer, also initially at the extremities.

The effect of these differentials is to generate stress distributions within a deck. The significance of these effects is largely related to the structural form with deep, solid slabs experiencing the largest stress differentials, potentially adding additional stresses to steel reinforcement and to the extreme fibres of the concrete sections.

Again, as temperatures resulting from climate change start to increase, the effect of differential temperature distributions could become more significant. Although highly unlikely to lead to a failure in isolation, when added to other effects, extreme differential temperature distributions could trigger, at the very least, a precursor event.

10.4.3   Thermal shock

A thermal shock occurs when there are sudden temperature changes in a very large bridge element, producing an even greater stress distribution through the section than might have been determined using standard differential temperature profiles. Significant crack propagation can result, requiring immediate repair interventions (Woyciechowski *et al.*, 2020).

In the context of climate change impacts, perhaps the most likely scenario might be in high summer, with a bridge experiencing very high temperatures to be suddenly and rapidly cooled by torrential rainfall. Although only very few instances of thermal shock have been identified, it is a phenomenon which could become more frequent if climate change predictions prove to be accurate.

10.5.   Wind

Extreme wind loading is an obvious resultant of the ever-increasing impact of climate change on the world's bridges. Irrespective of typical wind speeds which might be expected in a hurricane, typhoon or cyclone (all similar phenomena, but with names dependent on their geographic location), it is the frequency and, above all, the scale of wind gusts which will have the greatest consequences.

In terms of bridge management, most long-span structures should have procedures in place such that action can be taken when wind speeds and/or gust frequencies meet a predetermined threshold. Although national forecasting is becoming ever more reliable, on-site anemometers on the bridge providing a constant readout should be preferred. As well as speed, wind direction is also a factor which might influence decisions on precautionary measures. There are usually a variety of options to be considered, ranging from the banning of high-sided vehicles, through to lane closures to, *in extremis*, the full closure of the bridge. Dealing with wind will be different for almost every bridge; structural form and adjacent topography will both be significant. As an example of the former, a suspension or cable-stayed bridge with wind direction normal to the carriageway will have wind shadows at tower locations which will present problems to drivers. While this may not necessarily be of concern to the structure, it can lead to road traffic accidents and consequent delays.

An option used on many newer crossings is to employ wind shielding; one of the first bridges to use this solution was the Second Severn Crossing between Wales and England in the UK. Opened in 1996, and now known as the Prince of Wales Bridge, this 5128-m-long crossing has 2-m-high open steel fences in addition to vehicle restraint systems (severnbridges.org, 2024). The wind shielding was successful until Storm Eunice hit the UK on 18th February 2022 and the bridge had to be closed for the first time in its history. Of course, this innovation was part of the original design and the subject of rigorous aerodynamic testing. Retrofitting wind shielding will be problematic, as it will probably have a deleterious effect on an existing bridge's aerodynamic behaviour.

A case study of how wind gusts can have severe consequences occurred in New Zealand on 18th September 2020. Opened in 1959, and carrying a four-lane dual carriageway, the Auckland Harbour Bridge was widened ten years later with two additional lanes cantilevered from each side. The overall length is 1020 m with a main span of 243.8 m. The original form of the approach spans is an underslung truss which becomes a through truss at the main spans. The cantilevered sections are steel box girders. When a wind gust of 127 km/h struck the bridge in September 2020, two trucks on the main span were blown over. One of them hit the truss elements causing significant damage to one of the diagonals, prompting an urgent closure for several days pending repair. Although this was described as a 'freak' gust (NIWA, 2021), the New Zealand Transport Authority commissioned a detailed study into the event and to determine what measures were needed to improve their understanding of future risks and to improve their forecasting (Waka Kotahi, 2023).

10.6.   Sea level rise

Although in the short to medium term, sea level rise is not going to be catastrophically excessive (up to 0.3 m by 2050 and up to 0.6 m by 2100 (climate.gov, 2022)), it will need to be addressed in terms of adaptation for some low-lying bridges. Such adaptation will also need to address combinations with other climatic events, such intense areas of low pressure locally 'lifting' water levels and storm surges caused by high winds, and even the impact of tsunamis in areas of high earthquake risk.

Among the most vulnerable details are those foundation elements perhaps already in an intertidal zone. Of these, it is perhaps suspension bridge anchorage chambers, many of which already sit at or below water level, that are at most risk.

Another aspect here concerns shipping clearance. As well as sea level rise, the other driver in ensuring that bridges are able to be fully functional, and able to take the world's largest vessels, is commercial. The viability of ports whose access is limited by a major bridge, against a backdrop of ever-increasing shipping tonnages, is essential with respect to both local and national economies.

Probably the most dramatic case study in this regard concerns the Bayonne Bridge over the Kill Van Kull waterway between Staten Island, a Borough of New York City, and New Jersey, USA. Designed by Othmar Ammann and completed in 1931, when opened this was the longest steel arch bridge in the world with a main span of over 510 m and a clearance above water level of 46 m (Rastorfer, 2000). By 2016, however, the Panama Canal had been widened, allowing the next generation of shipping, the so-called Panamax cargo vessels, to more easily access the eastern seaboard of the USA. Well before this, however, the Port Authority of New York and New Jersey had commissioned a study to consider the raising of Bayonne's clearance by 20 m, an increase of over 43 per cent. Arguably at the expense of the original design's aesthetics, the project was completed in 2019 (WSP, 2019).

10.7.   Humidity

Although a direct link between higher levels of humidity and climate change may only be tenuous, it is a phenomenon which can be added to a list of other contributory factors potentially leading to a bridge failure or collapse. This will be especially the case in enclosed spaces in bridges, such as inside box girders, both in a marine environment and in one which will also experience high ambient temperatures.

A case study on how high levels of humidity may have been instrumental in a bridge collapse is that of the Koror-Babelthuap Bridge in Palau, an island nation in the western Pacific Ocean. Built in 1977, this was a three-span variable-depth prestressed concrete box-girder bridge with a main span of over 240 m, the longest of its kind in the world at that time. It collapsed without warning on 26th September 1996, claiming two lives. The collapse occurred just six weeks after the completion of remedial works which had sought to address some serviceability issues, notably excessive creep deflections. Frustratingly, no formal report on the collapse has ever been published as all the parties concerned had signed confidentiality agreements (Burgoyne and Scantlebury, 2008).

Later research, however, focused attention on the deflections caused by creep and the fact that the high levels of humidity within the box had not only accelerated corrosion of the post-tensioning strands but had also meant that prestress losses due to creep had been underestimated (Bažant *et al.*, 2012).

There will be many other bridges in which levels of high humidity in enclosed spaces could be an issue. As well as the insides of box-girders, elements such as suspension bridge anchorage chambers may be equally vulnerable. In either of these, lowering humidity levels through the use of retrofitted dehumidification systems is now recognised as a very effective adaptation practice.

10.8.   Ice accretion
Having addressed several phenomena relating to high temperatures, it must also be remembered that prolonged spells of very cold weather can also be attributable to climate change. As well as global temperature effects on bridges, another action which is already fairly common is that of ice accretion.

Of course, ice accretion is not just a bridge problem. Of probably the greatest concern is aircraft icing which is potentially disastrous as the additional ice layer can have a dramatic impact on the lift capacity of an aerofoil section of a wing (ScienceDirect, 2015).

Although ice accretion can hit any element of any structure, the primary concern lies with bridges which in themselves have an issue of aerodynamic performance, such as long-span suspension or cable-stayed bridges with aerofoil section box-girder decks. Those structural forms also have a feature of bridge elements above live carriageways, either in the form of overhead cable systems or cross members between towers. One such example is the Port Mann Bridge in British Columbia, Canada. This cable-stayed bridge opened in July 2012, but by December that year it had been plagued by ice accretion on the stay cables. This led to what the local press described as 'ice bombs' falling onto traffic lanes below. The problem was relatively easily addressed, however, with an immediate fix of a series of heavy collars fitted to each of the stay cables at their connection with the pylons (Meiszner, 2013). As snow and ice start to accumulate, a collar is released and falls by way of gravity, clearing the cables in the process.

Similar ice accretion problems have also been a concern in the UK at the Severn bridges (New Civil Engineer, 2009) and the Queensferry Crossing (BEAR Scotland, 2024).

10.9.   Sustainability and resilience
Underpinning nearly all of the climatic risks in this chapter is the need to make all transport networks, and bridge stocks in particular, both sustainable and resilient (Mitoulis *et al.*, 2022). Each of these aspirations should be a key part of proactive bridge management and, as has been covered elsewhere in this book, this requires understanding of both condition and risk, with that bridge manager's 'constant low level of anxiety' noted in Chapter 1 (Harvey, 2021). Both sustainability and resilience also demand an eyes-wide-open, vigilant approach, a questioning mind, and a desire for innovation.

The most important aspect of sustainability is longevity, making our bridges last as long as possible, safely squeezing as much life out of them for at least the duration of their design lives but preferably longer. This in turn, requires not just sound bridge management but also an improved understanding of the environmental pressures that those bridges will have to face in the coming decades. Masonry arch bridges are the UK's most common structural form and, although many are already well over 150 years old, there should be no complacency. Indeed, there is an imperative to make them even more sustainable (Steele *et al.*, 2003).

10.10. Future-proofing
Despite all best endeavours to embrace sustainability and resilience in an existing bridge stock, the time will come when a bridge needs to be replaced. Similarly, there is still a demand for new roads and railways around the world. With both scenarios, and especially with respect to climate change, it is essential to ensure that new designs not only meet current requirements but also those which might need to be extrapolated to be able to cope with future demands.

Perhaps the most obvious is waterway capacity. While a culvert or bridge span may be suitable to meet a current 1 in 100-year flood event, will this still be the case in 50 or 100 years' time? Although future-proofing in this sense could be hard to justify in terms of purely financial costs, here is the need for a much more sophisticated whole-life costing approach to be taken, and for clients to accept that they must plan for the longer term, including potentially unthinkable scenarios.

At the time of writing, there is no formal standard with respect to future-proofing, but it is hoped that the UK's technical approval process standard (National Highways, 2021) will add this requirement at its next revision.

10.11. Conclusion
Although it might be argued that the various climate-driven effects in this chapter are not new, it is their frequencies and intensities that must be recognised as being very different from those of, say, the 1990s.

Irrespective of whether climate change is human-made (and my personal conclusion is that there is little doubt that it is), bridge managers must recognise that 'times they are a changing'! No longer can we assume that the effects we have had to deal with for the last few decades will continue to be the norm in the immediate future.

REFERENCES

Bažant Z, Yu Q and Guang-Hua L (2012) Excessive long-time deflections of prestressed box girders. 1: Record-span bridge in Palau and other paradigms. *Journal of Structural Engineering* **138**(6): 676–686.

BEAR Scotland (2024) Queensferry Crossing ice accretion. https://www.bearscot.com/winter-portal-old/queensferry-crossing-ice-sensors-2/ (accessed 22/06/2024).

Burgoyne C and Scantlebury R (2008) Lessons learned from the bridge collapse in Palau. *Proceedings of the Institution of Civil Engineers – Civil Engineering* **161**(6): 28–34.

CAT (2024) Climate Action Tracker. https://climateactiontracker.org/ (accessed 11/06/2024).

CCRA (2022) UK Climate Change Risk Assessment 2022 – GOV.UK. https://www.gov.uk/government/publications/uk-climate-change-risk-assessment-2022 (accessed 10/06/2024).

Chatta H (2022) *Climate Crisis: Hassanabad Bridge Collapse Hunza.* Social Protection Resource Centre. https://www.sprc.org.pk/climate-crisis-hassanabad-bridge-collapse-hunza-pakistan-2022/ (accessed 16/06/2024).

climate.gov (2022) Climate Change: Global Sea Level. NOAA, Climate.gov. https://www.climate.gov/news-features/understanding-climate/climate-change-global-sea-level (accessed 21/06/2024).

climate.gov (2024) Global Climate Dashboard. NOAA, Climate.gov. https://www.climate.gov/climatedashboard (accessed 11/06/2024).

Dentith K (2022) Chapter 11: Scour. In *Highway Bridge Management* (Cole G and Fish RJ (eds)) ICE Publishing, London, UK, pp. 151–166.

Dora J and Ferranti E (2024) Infrastructure resilience under a changing climate: The urgent need for engineers to act. *Proceedings of the Institution of Civil Engineers – Civil Engineering* **177(5)**: 59–64.

Ebrahimi M, Djordjević S, Panici D and Tabor G (2020) A method for evaluating local scour depth at bridge piers due to debris accumulation. *Proceedings of the Institution of Civil Engineers – Bridge Engineering* **173(2)**: 86–99. 10.1680/jbren.19.00045.

England G *et al.* (2000) *Integral Bridges: A Fundamental Approach to the Time-Temperature Loading Problem.* Thomas Telford, London, UK.

Francis N (2021) Civil engineers' role in saving the world: Updating the moral basis of the profession. *Proceedings of the Institution of Civil Engineers – Civil Engineering* **174(5)**: 3–9.

Guggenheim D (dir.) (2006) *An Inconvenient Truth.* Film. Producers Laurie D, Bender L and Burns S.

Harvey W (2021) Some thoughts on competence. *Sixty-sixth Meeting of the Bridge Owners' Forum* (January 2021). https://www.bridgeforum.org/bof/meetings/bof66/BOF66%20-%20Harvey%20-%20Competence%20note-202101a.pdf (accessed 23/06/2024)

Hewson T, Ashoor A, Boussetta S *et al.* (2023) Medicane Daniel: An extraordinary cyclone with devastating impacts. ECMWF. https://www.ecmwf.int/en/newsletter/179/earth-system-science/medicane-daniel-extraordinary-cyclone-devastating-impacts. (accessed 11/06/2024).

Hulet KM, Smith CC and Gilbert M (2006) Load-carrying capacity of flooded masonry arch bridges. *Proceedings of the Institution of Civil Engineers – Bridge Engineering* **159(3)**: 97–103.

IPCC (2023) Climate Change Synthesis Report. https://www.ipcc.ch/report/ar6/syr/downloads/report/IPCC_AR6_SYR_LongerReport.pdf (accessed 10/06/2024).

legislation.gov.uk (2008) Climate Change Act 2008. https://www.legislation.gov.uk/ukpga/2008/27/contents (accessed 10/06/2024).

Meiszner P (2013) Port Mann Bridge 'ice bomb prevention system' debuts during today's snowy weather. *BC, Globalnews.ca.* https://globalnews.ca/news/1044434/port-mann-bridge-ice-bomb-prevention-system-debuts-during-todays-snowy-weather/ (accessed 22/06/2024).

Met Office (2022) UK climate projections: Headline findings. https://www.metoffice.gov.uk/binaries/content/assets/metofficegovuk/pdf/research/ukcp/ukcp18_headline_findings_v4_aug22.pdf (accessed 10/06/2024).

Met Office (2023) 2023: The warmest year on record globally. https://www.metoffice.gov.uk/about-us/news-and-media/media-centre/weather-and-climate-news/2024/2023-the-warmest-year-on-record-globally (accessed 10/06/2024).

Mitoulis SA, Domaneschi M, Cimellaro GP *et al.* (2022) Bridge and transport network resilience – a perspective. *Proceedings of the Institution of Civil Engineers – Bridge Engineering* **175(3)**: 138–149.

Mulhern O (2020) *The Time Lag of Climate Change.* Earth.Org. https://earth.org/data_visualization/the-time-lag-of-climate-change/ (accessed 23/06/2024).

National Highways (2021) CG 300 Technical Approval of Highway Structures. *Design Manual for Roads and Bridges.* National Highways, Birmingham, UK.

National Highways (2024) CS 469 Management of Scour and Other Hydraulic Actions at Highway Structures. *Design Manual for Roads and Bridges.* National Highways, Birmingham, UK.

Nepomuceno D, Webb G, Bennetts J, Tryfonas T and Vardanega P (2022) Thermal monitoring of a concrete bridge in London, UK. *Proceedings of the Institution of Civil Engineers – Bridge Engineering* **175(1)**: 16–34.

New Civil Engineer (2009) Falling ice strikes Severn bridge cars. *New Civil Engineer.* https://www.newcivilengineer.com/archive/falling-ice-strikes-severn-bridge-cars-12-02-2009/ (accessed 22/06/2024).

NIWA (2021) Scientists examine high winds on Auckland Harbour Bridge. https://niwa.co.nz/news/scientists-examine-high-winds-auckland-harbour-bridge (accessed 21/06/2024).

NOAA (2024) National Oceanic and Atmospheric Administration. https://www.noaa.gov/ (accessed 11/06/2024).

Obama B (2014) *Remarks by the President at the UN Climate Change Summit.* Office of the Press Secretary, The White House. https://obamawhitehouse.archives.gov/the-press-office/2014/09/23/remarks-president-un-climate-change-summit (accessed 09/06/2024).

Rastorfer D (2000) *Six Bridges: The Legacy of Othmar H Ammann.* Yale University Press, New Haven, USA.

ScienceDirect (2015) Accretion of ice – An overview. ScienceDirect Topics. https://www.sciencedirect.com/topics/engineering/accretion-of-ice#:~:text=The%20two%20stages%20of%20ice,mass%20on%20the%20blade%20surfaces. (accessed 22/06/2024).

severnbridges.org (2024) Other Challenges. https://severnbridges.org/other-challenges/ (accessed 21/06/2024).

Steele K, Cole G, Parke G, Clarke B and Harding J (2003) Environmental impact of brick arch bridge management. *Proceedings of the Institution of Civil Engineers – Structures and Buildings* **156**: 273–281.

United Nations (2015) Sustainable Development Goals. https://www.un.org/sustainabledevelopment/sustainable-development-goals/ (accessed 10/06.2024). Note: The content of this publication has not been approved by the United Nations and does not reflect the views of the United Nations or its officials or Member States.

United Nations (2016) The Paris Agreement. UNFCCC. https://unfccc.int/process-and-meetings/the-paris-agreement (accessed 09/06/2024).

Waka Kotahi (2023) Auckland Harbour Bridge: Desktop risk assessment of wind related vehicle incidents – Mar. 2023. https://www.nzta.govt.nz/assets/projects/auckland-harbour-bridge/AHB-desktop-risk-assessment-of-wind-related-vehicle-incidents-report.pdf (accessed 21/06/2024).

Wallace-Wells D (2019) *The Uninhabitable Earth – A Story of the Future.* Penguin Random House, London, UK.

World Meteorological Organization (2024). https://wmo.int/media/news/devastating-rainfall-hits-spain-yet-another-flood-related-disaster (accessed 27/11/2024).

Woyciechowski P, Łukowski P and Adamczewski G (2020) Thermal shock as a cause of cracking in concrete in massive bridge support elements – a case study. *Journal of Road and Bridge Research Institute,* Warsaw, Poland. https://www.infona.pl/resource/bwmeta1.element.baztech-6667821f-9236-4db8-99d0-836a3aa4aaad/tab/summary (accessed 20/06/2024).

WSP (2019) Bayonne Bridge raising opens NJ ports to world's largest ships. https://www.wsp.com/en-us/insights/bayonne-bridge-raising-opens-ports-to-worlds-largest-ships (accessed 21/06/2024).

Richard Fish
ISBN 978-1-83608-559-1
https://doi.org/10.1108/978-1-83608-556-020251012
Emerald Publishing Limited: All rights reserved

Chapter 11
The shape of things to come?

Although it may seem presumptuous to entitle this chapter after a classic science fiction novel (Wells, 1933), HG Wells described that book as the 'history of the future'. While this chapter cannot compete with that strapline, it can perhaps give a glimpse into the opportunities that may present themselves, either in earlier research which has since become commonplace, or in some ideas which may yet prove to be transformative. Earlier chapters have considered what we should be doing in the immediate term to raise awareness of our professional responsibilities to keep safe the travelling public, but when it comes to bridge management in the *twenty-second* century, might we find the bridge manager sitting in an air-conditioned underground control room bunker (possibly because the heat outside is too intense?) watching a bank of screens and sending out drones to inspect their bridges and robots to fix them? Is such a scenario utopia or dystopia? To misquote George Orwell[1]: if there is hope, does it lie in technology? Possibly; to be discussed.

11.1.   Health warning

Parts of this chapter, more than any other in this book, could carry the jeopardy of possibly being outdated by the time of publication. It should be taken, therefore, as just a snapshot of not only the position in mid-2024 but also my views at that time.

11.2.   Structural health monitoring (SHM) or structural performance monitoring (SPM)?

Firstly, while the acronym SHM, and the phrase 'structural health monitoring', is now in common parlance among bridge engineers, I have long believed that this gives the impression of a sick patient lying in a hospital bed and wired up to numerous monitors that are all delivering information on life support systems. My preference has always been that this should be 'structural performance monitoring'. In other words, rather than a vision of a sickly patient, we should be making comparisons with an athlete at the peak of their form, always hoping to improve on their personal bests. Despite presenting on this idea at numerous conferences (Fish, 2016), it seems to have fallen on deaf ears because SHM remains the abbreviation that seems to have become embedded in the bridge engineer's lexicon and one which everyone now appears to be comfortable with.

11.3.   Structural health monitoring

Accepting the term, there are many developing strands in SHM which are too many to be covered in these pages (Catbas and Avci, 2023; Anderson and Cousins, 2022). There is also the parallel question of aligning SHM with the principles of risk management as described in Chapter 8 (Colford *et al.*, 2024).

---

[1]The actual quote from Orwell's dystopian novel, *1984*, is, 'If there is hope, it lies in the proles'

The obvious question for SHM is simply this: is this not just effective bridge management? In a sense it certainly is. As we have seen, there are no more important tasks in the bridge management process than inspections. As has always been the case, inspection reports provide the base data from which all other bridge management activities and decisions are determined and actioned. If this somewhat clunky, even analogue, process is already providing sufficient information to enable a stock to be managed effectively, is there a need for anything more sophisticated? It is very difficult to determine exact percentages, but I suggest that in about 99 per cent of the world's bridges, probably not. For the other 1 per cent, and these are generally the major, long-span crossings, there is a definite merit in using remotely accessible sensors to provide a continuous stream of data to inform those decisions.

Of course, there is a grey area in which a traditional visual inspection regime will take advantage of some basic systems to enhance objective measurement. Obvious among these will be crack width tell-tale gauges which, although they may not be accessible remotely, can easily be observed at a greater frequency than every two years.

Returning to the 1 per cent, however, these are likely to be bridges of greater complexity, most of which will have been built in the last few decades and which will have had an array of sensors embedded in their design and construction from the outset. A relatively smaller number will be older bridges which have been retrofitted with sensors with specific purposes, intended to target explicit concerns. Occasionally, both are employed in a complementary manner in parallel structures, one old and one new (Angus, 2019).

Structural health, therefore, need not simply be reliant on a somewhat subjective assessment of a bridge's condition every two years. The employment of sensors gives the opportunity for real-time information. Although this may be extremely appealing in terms of being able to access streams of data, the DIKW principle in Figure 4.1 is highly relevant here. Post-processing the data into useful information is an essential requisite in ensuring that the goals of knowledge and wisdom can be attained.

The details of the exact techniques of monitoring are far too numerous to be recorded here, but an excellent summary of current best practice in the UK, including numerous case studies, is already available (Middleton *et al.*, 2016). Similarly, other references give an international perspective (Spuler *et al.*, 2011) but all tend to use similar technologies.

11.4. Data gathering

Traditionally, visual inspections have been relied upon to provide the various pieces of data which can inform bridge managers about the condition of a bridge (Phares, 2005). This can be supplemented by intrusive investigation and sampling of materials to add extra layers to the understanding of condition and behaviour.

For many bridge inspections, just getting near enough for long enough can be a serious logistical problem. For a Principal Inspection (or Detailed Examination as it is known for UK rail bridges), costs of traffic management, access equipment, moveable plant and lighting are not inconsiderable. Demands for minimal interruption of transport networks already dictate that inspections of bridges over motorways and trunk roads are regularly undertaken at night, often requiring detailed planning akin to a military operation. Additional factors, such as less than perfect lighting, time pressures towards the end of the inspection period before the road must be fully reopened, and just plain tiredness, can diminish the effectiveness of night-time inspections (Laing, 2022).

Even more problematic are inspections over railways (for highway and railway bridge owners alike) which have to be planned many months in advance with prebooked rail possessions, which themselves can be cancelled at very short notice. Attempting to squeeze as much capacity as possible out of the UK's rail networks has already led to political calls for a '24-hour railway' (BBC, 2018), theoretically leaving no time whatsoever for inspection and maintenance, with potentially dire consequences.

How then can we supplement traditional inspections with new technology? For this fundamental aspect of bridge management, some relatively recent technological advances are covered in the following sections.

11.5. Photogrammetry

Photogrammetry is the art and the science of patching together several hundred digital photographs to provide both a detailed geometry and a high-resolution surface condition to produce an accurate 3-D model (Harvey and Harvey, 2016). Once the model has been created, consequent site visits can repeat the photographic exercise so that comparisons may be made to determine any changes in deformations or visible defects. Although the primary subject of the reference is masonry arch bridges, the principle can be applied to other bridge types.

11.6. LiDAR or laser scanning

An extension to photogrammetry is Light Detection and Ranging (LiDAR) which uses lasers to accurately measure dimensions and therefore create a 3-D model of a bridge. LiDAR has many more applications than just in bridge inspections and has been used as a tool for several years to create aerial maps for any type of earth science application (Cracknell et al., 2007). As an aside, this is also the basic technology being used in driverless car development.

Returning to bridge-specific uses, laser scanning has been proved to be a useful tool in taking some of the subjectivity out of bridge inspections and, therefore, the data that they provide (Tang et al., 2007). That said, bridge condition is based on both severity and extent of defects which, at the moment, remains reliant on the bridge inspector's judgement.

11.7. Unmanned aerial vehicles

In addition to photogrammetry or laser scanning, employing unmanned aerial vehicles (UAVs, or more commonly known as drones) for bridge inspection would seem to be a logical move, doing away with many of the problems of traffic management, access equipment and so on. The use of UAVs in almost every sector, including recreational use, has grown at an unprecedented rate since the turn of the century. And yet there still seems to be some reluctance to fully embrace this technology for bridge inspections. The arguments against UAVs include: can they distinguish a cobweb from a crack? Or, if only they could hit a concrete surface to test for delamination or masonry to test for drumminess (as would be required for a Principal Inspection in the UK).

The use of UAV inspections is rapidly becoming seen as being complementary to more traditional hands-on inspections in which every part of a bridge should be physically accessible within touching distance. An initial UAV fly-by can easily identify hot spots which might need closer attention from a physical inspection, thereby efficiently targeting key areas which will need that hands-on contact.

The next step will be to combine UAV inspections with Artificial Intelligence (AI) or machine learning (Perry *et al.*, 2020). These topics are covered below.

11.8. Artificial intelligence (AI) and machine learning

Although these terms are often assumed to be one and the same, AI is a much broader subject with applications being developed in every sector. Machine learning is a subset of AI, referring specifically to computers which can perform tasks 'without being explicitly programmed'. It is also not the case that these have been twenty-first-century developments. In fact, in was an IBM scientist, Arthur Samuel, who first used these terms as early as 1959 (Koza *et al.*, 1996).

AI can be used throughout a bridge's life cycle, from assisting with fundamental early decisions on whether a bridge is needed in the first place, through to design and construction where it can have a role in optimising materials such that the bridge is working as efficiently as possible and minimising embedded carbon (Reich, 1996). Bearing in mind the emphasis of this book, however, its applications discussed here will be limited to operation, maintenance and management.

With regard to these aspects, it must be assumed that there is sufficient data being made available for the technology to analyse. For the most part, certainly in terms of bridge operation, that data will be provided by a series of sensors either built in during construction or added to an existing bridge targeting a specific purpose. Examples might include anticipating site-specific weather conditions, unusual traffic patterns or loads, or changes in dynamic behaviour. The last of these has been found to be effective even for very stiff bridges and, when linked to satellite technology (see below), may even be an early indicator of possible scour hole development (Selvakumaran *et al.*, 2018). The specific case study in this reference relates to the partial collapse of the seventeenth-century multispan masonry arch Tadcaster Bridge over the River Wharfe in Tadcaster, UK, in December 2015. The bridge had been a test bed for the satellite monitoring project for two years prior to the collapse. Unfortunate as the collapse was, it did at least prove that movements were sufficient to provide a degree of early warning for such high-risk bridges in the future.

Turning to bridge maintenance and management, and returning to inspections, AI can assist with the programming of inspections, notably Principal Inspections for highway bridges in the UK where a risk-based relaxation of the usual six-year frequency can allow an extension of up to 12 years (National Highways, 2020). Similarly, bridges in poor condition, and being monitored as substandard structures, may well require a much more frequent inspection regime. AI or machine learning can also greatly assist with the interpretation of images being collected by way of UAV flybys, especially when comparing those derived from a series of inspections over a number of years.

AI can also be used to evaluate maintenance interventions and their prioritisation, potentially conducting small-scale risk analyses as part of the process. Other specific examples may include fatigue analysis, including helping to determine exactly where an element, or a bridge, might be on the S-n curve[2] at any point in time.

Above all, and applicable to all of the above, AI can be used to detect human errors or oversights throughout the bridge management processes. As has been shown in earlier chapters, it is the

---

[2]Also known as a Woehler curve, this is a plot (on a logarithmic scale) of the magnitude of alternating stress levels against the number of cycles to failure.

human factor which is a significant contributory cause in many failures and collapses. Having an analytical and objective check by a nonhuman member of the team may well prove to give the greatest benefit from this technology.

11.9. Data management

A recent survey on the use of data in bridge asset management by bridge owners and managers found that there is a significant variation across the sector. The survey results, however, recognised that there was an opportunity to develop systems that could lead to the development of 'smart' bridge management applications (Bennetts *et al.*, 2020).

As well as providing a look into the future, this reference also gives a snapshot of the current situation based on interviews with bridge practitioners. In the round, it seems that, although most recognise the benefits of a system based on smart data, AI and machine learning, there remained some scepticism as to how soon this could happen, how effective it might prove to be, and how much it would cost for a typical stock. Despite these views, it seems that a move to intelligent bridge asset management is within sight. A move which *should* see the trend of rising numbers of collapses at least being stemmed and hopefully reversed.

11.10. Digital twins

Closely linked to the advancement towards better use of data and intelligent asset management is the concept of digital twins. This subject can draw together many of the above topics with an outcome that there are two identical twin bridges – both three-dimensional but one is 'real' and the other virtual, a digital model correct in every detail (Deh Bozorgi, 2024).

Building the virtual twin is best achieved from the time of the actual bridge's design and construction, utilising works information and site records and normally supplemented by embedded sensors. The digital model can then be calibrated against the real bridge in terms of its performance under external actions such as live load and weather effects.

Although the digital twin concept can also be applied to existing bridges, calibration of the two will require some retrofitted sensors, potentially needing to be supplemented by load testing to compare actual and theoretical structural behaviour (Jasiński *et al.*, 2023).

11.11. The internet of things (IoT) and sensors

The IoT is a concept in which a network of connected devices (each with some sort of sensor within it) can exchange data with one another and the cloud, thereby forming a network which can transfer data to its most applicable receptor. Here AI can also help to analyse all incoming data to produce outputs in a form which will be the most helpful (Yasar and Gillis, 2024).

Although the IoT can have multiple applications across the digital world, it fits neatly with the use of sensors on bridges and the digital twin concept discussed above. In terms of collecting data, it has even been suggested that smartphones (now with extremely sophisticated sensors within them) in cars using a highway network can be used to contribute to the IoT to help monitor a bridge stock (Matarazzo *et al.*, 2022).

The obvious risk with reliance on the IoT and data storage is one of security, either from malevolent hacking or widespread outages. Although data backups may be available, it is not unlikely that accessibility issues will require some significant contingency planning.

11.12. Satellites

Mention has already been made above, in the context of the partial collapse of the Tadcaster Bridge in the UK in 2015, of the potential for satellite technology to be able to monitor changes in dynamic behaviour during the formation of a scour hole.

Too numerous to note here, there are many more research projects underway or applications being developed in this sector at the time of writing. Among these, however, the Italian Space Agency has been working with the Rochester Bridge Trust in the UK to detect displacements on one of the Trust's bridges (Gagliardi *et al.*, 2023) and the UK Satellite Application Catapult funding programme has also been supporting a project to monitor unexpected bridge displacements (Cusson *et al.*, 2021).

11.13. Robotics

The majority of this chapter has focused on the bridge *management* process but are there applications which might be able to deliver *maintenance* intervention projects? Although not yet in common use, the answer to this question is yes, with various applications.

Firstly, with respect to cable-supported structures, the cable-climbing robot is almost upon us (Xu *et al.*, 2021). It seems likely that by 2030, such a robotic device will be able to both inspect and maintain cables up to 140 mm in diameter. Maintenance could include repainting or even minor repairs to uPVC sleeves.

Another application applies to painted steel bridges and a robotic system which could be used to detect deterioration in protective systems and undertake repairs. These might include removing old paint and corrosion products, surface preparation and applying a localised repair (Liu *et al.*, 2008).

This is an area in which further progress is likely to be made in the subsequent decades. If one considers the transition in car assembly lines from humans to machines in the last few decades, it would seem likely that there could be a similar growth in the bridge maintenance sector. It is probable that routine maintenance, such as touching up localised paint defects, will be the low-hanging fruit but it would seem churlish to impose limits on future potential opportunities at this time. The test will be in the reliability of robots to do everything that is expected. Linked with AI, however, it seems likely that this technology could help to prevent failures in the round.

11.14. Conclusion

It seems almost certain that all of this rapidly emerging technology could deliver a considerable benefit to the cause of preventing bridge collapses and failures in the not-too-distant future. The question remains, however, as to whether all of the above innovations can be used throughout an existing bridge stock. While they may be best suited to new, long-span bridges, and these may be the structures which can deliver the better ratios of benefits against cost, is it economic to think that they can be applied to all new bridges, irrespective of span? And what of the overwhelming majority, those existing bridges of multiple form, material and size which are probably at greatest risk of failure? While it may seem possible that all bridges can be beneficiaries, there needs to be a reality check here.

The opening paragraph of this chapter ended with the question: if there is hope, does it lie in technology? The answer from all of the above text must be a yes, at least in part. However, the scale of the problem of the ageing post-war bridge stock, and the sparsity of resources to properly manage it, will mean that the current rate of failures is likely to remain little changed in the short to medium term. An over-reliance on new technology is unlikely to completely turn the tide of bridge collapses but its pragmatic application, targeted at the highest-risk bridges, may at least help to reduce their number. Having started with a science fiction classic, I end with another: perhaps new technology is pointing the way to a brave new world?

REFERENCES

Anderson M and Cousins D (2022) Structural health monitoring. In *Highway Bridge Management.* (Cole G and Fish RJ (eds)). ICE Publishing, London, UK.

Angus E (2019) Making assets smart: Digitally enabled asset management on the Forth road bridges. *Bridge Owners Forum Meeting 61*, May 201. https://www.bridgeforum.org/?post_type= wpdmpro&p=581&wpdmdl=581&refresh=668316a00cb611719867040 (accessed 01/ 07/2024).

BBC (2018) Trains could soon run 24 hours a day across network – rail boss. https://www.bbc. co.uk/news/uk-43961628 (accessed 16/07/2024).

Bennetts J, Vardanega P, Taylor C and Denton S (2020) Survey of the use of data in UK bridge asset management. *Proceedings of the Institution of Civil Engineers – Bridge Engineering* **173(4)**: 211–222.

Catbas N and Avci O (2023) A review of the latest trends in bridge health monitoring. *Proceedings of the Institution of Civil Engineers – Bridge Engineering* **176(2)**: 76–91.

Colford BR, Zhou E and Pape T (2024) Structural health monitoring – a risk-based approach. *Proceedings of the Institution of Civil Engineers – Bridge Engineering* **177(2)**: 89–98.

Cracknell AP and Hayes L (2007) *Introduction to Remote Sensing*, 2nd edn. Taylor and Francis, London, UK.

Cusson D, Rossi C and Ozkan IF (2021) Early warning system for the detection of unexpected bridge displacements from radar satellite data. *Journal of Civil Structural Health Monitoring* **11(1)**: 189–204.

Deh Bozorgi S (2024) Digital twins and data strategy. *Transportation Professional*, The Chartered Institute of Highways and Transportation. March/April 2024, pp. 8–9.

Fish RJ (2016) Innovation in the design and management of suspension bridges – past, present and future. *Technical Program Papers, The 9th International Cable Supported Bridge Operators' Conference*, Halifax, Nova Scotia, Canada, June 2016, pp. 300–307.

Gagliardi V, Ciampoli L, D'Amico F *et al.* (2023) Bridge monitoring and assessment by high-resolution satellite remote sensing technologies. https://repository.uwl.ac.uk/id/eprint/7451/1/ Spie_Bridge monitoring and assessment.pdf (accessed 22/07/2024).

Harvey W and Harvey H (2016) 3D modelling – Bill Harvey Associates Limited. https://www.billhar-veyassociates.com/photogrammetry (accessed 16/07/2024).

Jasiński M, Łaziński P and Piotrowski D (2023) The concept of creating digital twin bridges using load tests. https://www.mdpi.com/1424-8220/23/17/7349 (accessed 22/07/2024).

Koza J, Bennett F, Andre D and Keane M (1996) Automated design of both the topology and sizing of analog electrical circuits using genetic programming. *Artificial Intelligence in Design '96*. Springer, Dordrecht, pp. 151–170.

Laing M (2022) Bridge inspection. In *Highway Bridge Management* (Cole G and Fish RJ (eds)). ICE Publishing, London, UK.

Liu D, Dissayanake G, Manamperi P and Brooks P (2008) A robotic system for steel bridge maintenance: research challenges and system design. https://www.researchgate.net/ publication/228715804_A_Robotic_System_for_Steel_Bridge_Maintenance_Research_ Challenges_and_System_Design (accessed 23/07/2024).

Matarazzo TJ, Kondor D, Eshkevari S *et al.* (2022) Crowdsourcing bridge dynamic monitoring with smartphone vehicle trips. *Nature* **1(1)**. https://www.nature.com/articles/s44172-022-00025-4 (accessed 22/07/2024).

Middleton C, Fidler P and Vardanega P (2016) *Bridge Monitoring: A Practical Guide.* ICE Publishing, London, UK.

National Highways (2020) CS 450 Inspection of Highway Structures. *Design Manual for Roads and Bridges.* National Highways, Brimingham, UK.

Perry BJ, Guo Y, Atadero R *et al.* (2020) Streamlined bridge inspection system utilizing unmanned aerial vehicles (UAVs) and machine learning. *Measurement* **164**:108048. https://www.sciencedirect.com/science/article/abs/pii/S0263224120305868 (accessed 17/07/2024).

Phares BM (2005) Visual inspection techniques for bridges and other transportation structures. In *Inspection and Monitoring Techniques for Bridge and Civil Structures.* (Fu G (ed)) Woodhead Publishing Ltd., Cambridge, UK.

Reich Y (1996) Artificial intelligence in bridge engineering microcomputers. *Civil Engineering* **11**: 433–445.

Selvakumaran S, Plank S, Geiß C *et al.* (2018) Remote monitoring to predict scour failure using interferometric synthetic aperture radar (InSAR) stacking techniques. *International Journal of Applied Earth Observation and Geoinformation* **73**: 463–470.

Spuler T, Moor G, Berger R and Watson R (2011) Remote structural monitoring systems for long-term confidence in a structure's condition. In *Modern Techniques in Bridge Engineering* (Mahmoud KM (ed)) Taylor and Francis, London, UK, pp. 255–263.

Tang P, Akinci B and Garrett J (2007) Laser Scanning for Bridge Inspection and Management. https://www.researchgate.net/publication/233686297_Laser_Scanning_for_Bridge_Inspection_and_Management (accessed 16/07/2024).

Wells HG (1933) *The Shape of Things to Come.* Penguin Classics, London, UK.

Xu F, Dai S, Jiang Q and Wang X (2021) Developing a climbing robot for repairing cables of cable-stayed bridges. *Automation in Construction* **129**: 103807. https://www.sciencedirect.com/science/article/abs/pii/S0926580521002582 (accessed 23/07/2024).

Yasar K and Gillis AS (2024) What is the Internet of Things (IoT)?. https://www.techtarget.com/iotagenda/definition/Internet-of-Things-IoT (accessed 22/07/2024).

Richard Fish
ISBN 978-1-83608-559-1
https://doi.org/10.1108/978-1-83608-556-020251013

Chapter 12
Summary and conclusions

Having approached the subject of bridge failures not only by looking back at historical collapses but also by assessing what must be done to prevent similar events in the future, or at least reduce their number, it is important to draw together the various strands to try to inform a national strategic plan (should one ever be prepared) which may go some way in achieving that goal.

12.1. Collapses and causes

Firstly, in Part 1, I have deliberately chosen to focus on only a relatively small sample of bridge collapses, compared to the total throughout history. My objective was never to simply provide a list of failures, not least as this has been very successfully achieved by others (Scheer, 2010). Although this reference provides a very helpful and comprehensive account, it does so on an almost statistical basis, possibly at the expense of the personal tragedies and emotional trauma that lie behind every victim and, most probably, every survivor. On this point, while I have generally provided numbers of casualties, both killed and injured, I have done so with some reservations as I would not want the significance of a collapse to be measured solely on such numbers. Some collapses have offered the wake-up call that urgent action was needed to avert repetition – for example, although the Ynys-y-gwâs collapse in 1985 was only of a modest span and claimed no casualties, it served to alert our profession to a potentially significant problem in post-tensioned bridges, not only in the UK but also overseas.

All of the historical case studies in Chapter 5, therefore, were chosen because they have yielded a net benefit to bridge engineering. Lessons have been learnt from each, which have improved our knowledge and understanding of many of the causes of collapses cited in Chapter 3; this has provided us with a better understanding of materials, improved design rules to cover effects such as aerodynamic behaviour and fatigue, and an improvement in the procedures and processes of design and construction.

The twenty-first-century collapses covered in Chapter 6 should also be providing similar lessons. Apart from the earliest, de la Concorde (2006) and I-35W (2007), as well as Nanfang'ao (2019), at the time of writing the investigations into the others have yet to be reported. While it is tempting to make some assumptions on the likely contributory factors and even causes, no real and transparent lesson sharing can happen until formal reports have been made publicly available.

As well as the case studies covered in some detail in Part 1, others have been inserted elsewhere into the text to support the subject matter of other chapters throughout the book. It is an interesting exercise, however, to judge which, if any, of the collapse case studies have had no 'people' issues as contributory factors.

12.2. The animal part of the machinery

Included in Chapter 3 is a quotation of John Smeaton's (Skempton, 1981) which includes the phrase used as the title of this section. Smeaton's implication is that the most significant risk in any civil engineering venture lies with people. This applies equally to bridge engineers but while mistakes made by designers and contractors may well continue, despite more rigorous processes and procedures being introduced, the emphasis in this concluding chapter must reflect on the role of the bridge manager and some of the challenges that we face as professional engineers in this specific sector.

Human factors in failures are not unique to bridges, nor to civil and structural engineering. The distinguished chemical engineer and internationally recognised proponent of safety matters, Professor Trevor Kletz, made the following comparison (Anderson, 2024).

> 'For a long time, people were saying that most accidents were due to human error and this is true in a sense, but it's not very helpful. It's a bit like saying that falls are due to gravity.'

Although human error may almost always be a contributory factor in any failure, it is important not to spend too much energy on apportioning blame. Even if the failure is found to be due to negligence or malevolence, deeper questions must be asked into organisational or systemic issues that had led an individual, or group of individuals, to this position. The UK's Chartered Institute of Ergonomics and Human Factors has covered these points in some detail, as well as emphasising not just the need to learn from failure but also pointing out that the learning from those lessons must be embedded in the culture of an organisation, in order for them to be truly effective in the long term (CIEHF, 2020).

As far as the case studies in Part 1 are concerned, the reader must judge the effectiveness of learning in each. For some of these studies the lessons learnt are undoubtedly invaluable. Others, less so, even to the extent that learning has, at best, been simply a gesture limited only to the close circle of those directly involved. Knowledge sharing must never be on a 'need to know' basis.

Looking forward then, and this is the point of Part 2, how can we seek to address some of the human factors in order not to see history repeating itself in terms of bridge failures and collapses? The following sections cover some key traits, within individuals, organisations and, indeed, our collective profession.

12.3. So many Cs

Although not my direct intent, and certainly not trying to emulate a management textbook, so many issues that need to be addressed to keep our bridges safe seem to start with the letter C. The following are all equally valid, either at a personal level, within bridge management organisations, or in our wider professional accountability.

Competence is an essential prerequisite for anyone doing a job, from artisans to professionals. Competence is fundamental in terms of the level of respect that members of the public should expect to have in professions such as lawyers, doctors and surgeons, and teachers. There should be no difference, therefore, in their expectations from bridge engineering professionals in whom they literally trust their lives when using a bridge.

Competence, however, must be tested and proven. Not just once, but continuously. This is **continual professional development (CPD)** and should be a key part of every engineer's career from initial qualification through to retirement. CPD should be relevant to the specific tasks that an individual is expected to perform, and it should also be recorded. Engineering institutions in the UK expect CPD to help identify gaps in knowledge. Taking the UK's Institution of Civil Engineers as an example (ICE, 2024), CPD is in two parts, firstly through a development action plan which looks forward (usually for the year ahead) and secondly through a professional development record, which, as the name suggests, looks back over the previous year.

As noted in Chapter 7, probably the most important of all bridge management activities is inspections. Competence of bridge inspectors is vital. It is concerning, therefore, that in many countries the inspectors, their inspections and their reporting are often not up to the required standard. There are several collapse or close-call case studies in this book which would have been preventable if trained, competent inspectors had been used, and an equally competent bridge management team had been available to challenge initial findings.

Closely linked to competence is an individual's **capability**. This is perhaps better defined as their degree of independent working and includes the need to have an awareness of one's own limitations. Asking for help or advice should be regarded as a sign of strength, not of weakness. It does, however, also require a sympathetic and understanding line manager, which in turn is usually dependent on organisational culture. Capability can also refer to a team as well as an individual and, here, a team's dynamic can be crucial, with trust being an implicit requirement.

Capacity can refer to anything from a single team to the number of competent, trained bridge professionals in a country. In simple terms, it means the number (and variety) of people needed to manage a bridge stock. The obvious question, however, is what number is the right number? While it is tempting to think that a simple benchmarking exercise, based on a ratio of, say, the total deck area of a stock to the number of people responsible for its management, might provide that answer, there are so many variables that this is no easy task. It is also not just a numbers game; on the people side, another metric is undoubtedly experience. And when cuts are imposed on public sector bridge owners, it is usually those nearest retirement who are made offers too good to refuse. In so doing, there is a risk that competence, capability and capacity of the bridge management team will dramatically reduce at a stroke.

Several case studies throughout the book have shown that poor **communication** has been a root cause of bridge failures. No matter how well qualified and how experienced individuals might be, good communication is an essential part of ensuring that a team is functioning effectively, efficiently and, above all, safely. Good communication can be complex. It must not just be top-down, command-and-control orders and instructions. It must also allow and even encourage suggestions on a bottom-up basis, as well as having the freedom to discuss options without the senior manager having to be party to all conversations. Possibly of even greater importance is the communication between designer and contractor, usually by way of their site staff. Again, case studies in these pages demonstrate how important this is and probably none more so than the 1970 West Gate Bridge collapse covered in Chapter 5.

The last two Cs are possible outcomes from all of the above. If we cannot be sure that we are not only aware of the levels of competence, capability, capacity and communication but also taking action to address them, are we not in danger of underestimating them and therefore at risk of **complacency**? Chapter 1 referred to the late Professor Bill Harvey's warning that 'any bridge engineer worth his salt should be living in a state of constant low-level anxiety'. Therefore, if we are not living in this state of low-level anxiety, is the corollary that we are being complacent? I would argue that the answer is yes; not least, as has been shown, that the condition of every country's bridge stock is continuing to deteriorate.

The last outcome is **confidence**, which may be applied in two ways. Are we as a profession confident that we are doing all we can to keep our bridges safe? Similarly, are the travelling public right to have the confidence in bridge managers that we are keeping them safe? Perhaps citizens of Italy (Polcevera), India (Morbi) and the USA (I-35W), for example, should not be quite as confident as those of other countries; at least for the time being.

12.4. The world's bridge stocks

Having addressed the current state of people issues, the other principal concern remains the condition of our ever-ageing bridge stocks. As we have seen, bridge design life (of 120 years in the UK) is an arbitrary number. The distribution of actual life against number of bridges will probably look something like the traditional standard distribution as shown in Figure 12.1.

This suggests that the greatest number of bridges will achieve the 120 years. Many will last considerably longer and the majority of these are likely to be masonry arches which, if appropriately managed and maintained, can last several times that number. Many others, however, will last a lot less than their design life, principally because of the various issues covered in earlier chapters as well as possible unknown unknowns that may yet come to light. This distribution, however, is only likely to be accurate if *all* bridges are being effectively managed and if maintenance interventions can be implemented as soon as they are needed. With due cognisance of the current position for most bridges (and most bridge owners) around the world, it is highly unlikely that this will happen. Factor in the additional pressures from climate change and the net effect, therefore, is that the distribution of numbers against age will be nearer to that shown by the dashed line in Figure 12.2.

Figure 12.1 Standard distribution of bridge numbers against age (author's own)

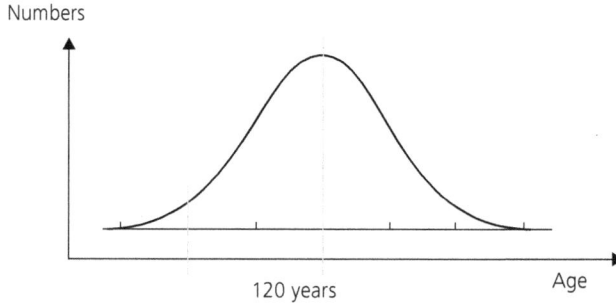

Figure 12.2 Indicative modified distribution of bridge numbers against age (author's own)

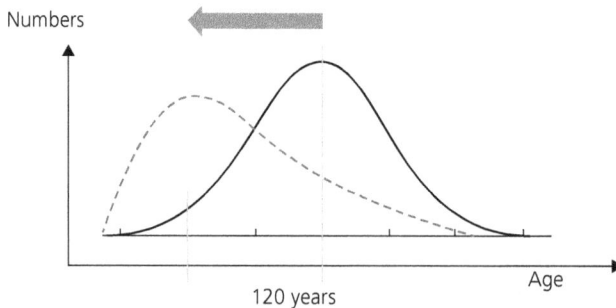

Although the area under the two curves is the same, this shows that the greater number of bridges are likely to last a lot less than the 120-year design life, and that many more are likely to need urgent attention or even replacement within the first few decades after commissioning.

12.5. The need for investment

Is the status quo with respect to bridge stocks tenable? This is obviously a rhetorical question because most of this book has sought to demonstrate the reality that things are likely to get much worse before they get any better.

Investment in our bridges, and in the primary driver of keeping safe the travelling public, must be twofold: money and people. The former should be hypothecated, with protected budgets based on needs, but, for publicly owned bridge assets, this is so much easier to say than to achieve. Politicians who control the purse strings must be lobbied and persuaded by bridge managers that there is a growing need for urgent action. Prevarication will simply increase the budgetary resources required year on year for firefighting, producing an ever-growing bow wave of problems such as bridge closures, restrictions and even collapses.

Investing in people, in the context of all of the points made in Section 12.3, is the other urgent necessity in terms of both numbers and quality. The competence and skill set of the whole team, from inspectors to decision-making bridge managers, must be non-negotiable.

Finally, even if budgets are protected and a stock is being effectively managed, the other metric relates to capacity and skills in the construction sector – those companies who will be asked to deliver the many hundreds of strengthening or major maintenance projects needed each year to drag the condition of the stock back to where it should be.

12.6. Brave new world?

Chapter 11 touched on the twenty-first-century technological revolution which may possibly make the bridge manager's life easier and our bridges safer. As noted, and as things stand at the time of writing, this may well be the case for long-span bridges, either through sensors embedded from construction or those retrofitted at a later date. The challenge, however, is to make this technology applicable to very large numbers of small- to medium-span bridges, as well as being both affordable and helpful in terms of the data and information provided. For the former, budgets for technological innovations will be competing with those for more mundane needs such as reactive basic maintenance demands. They are unlikely, therefore, to be transformative for the majority of our bridges.

As negative and unambitious as this conclusion might seem, I question whether my view is an outlier to that of other bridge management professionals. I would like to think that most would agree that new technology cannot provide all the answers to the problems they face on a day-by-day basis.

12.7. Conclusion

Completed in 2024, this book could be seen as a status report on bridge condition in the round, and on the risks of failures in particular. I have sought to hold a mirror up to the current situation to show, with the benefit of history, the degree of risk that we face and to question whether we are doing enough to mitigate that risk. I have also tried to extrapolate forwards to try to gauge what lies ahead. I have not set out to be prophetic but only time will tell the accuracy of my views.

It seems, however, that the demands on bridge managers will continue to grow to the point that they are constantly firefighting all of the immediate issues that they have to deal with. There is no quick and easy fix for our ageing bridge stocks but having a well-funded, planned, proactive maintenance regime seems to be the only way to keep them at least in a steady state.

At the moment, budgets and professional capacity determine maintenance. In an ideal world, professional capacity should not be the issue and maintenance needs will determine budgets. Or is this wishful thinking?

REFERENCES

Anderson M (2024) Remembering Trevor Kletz – Human Factors 101. https://humanfactors101. com/2020/10/31/remembering-trevor-kletz/ (accessed 11/07/2024).

CIEHF (2020) *Learning from Adverse Events*. Chartered Institute of Ergonomics and Human Factors. https://ergonomics.org.uk/resource/learning-from-adverse-events.html (accessed 11/07/2024).

ICE (2024) Continuing Professional Development (CPD) guidance. Institution of Civil Engineers. https://www.ice.org.uk/download-centre/continuing-professional-development-cpd-guidance (accessed 12/07/2024).

Scheer J (2010) *Failed Bridges: Case Studies, Causes and Consequences*. Ernst & Sohn, Berlin, Germany.

Skempton A (1981) *John Smeaton FRS*. Thomas Telford Ltd., London, UK.

Richard Fish
ISBN 978-1-83608-559-1
https://doi.org/10.1108/978-1-83608-556-020251014
Emerald Publishing Limited: All rights reserved

Index

www.ingramcontent.com/pod-product-compliance
Lightning Source LLC
Chambersburg PA
CBHW081106220326
41598CB00038B/7245